Automated Network Technology

Automated Network Technology

The Changing Boundaries of Expert Systems

Carl P. Catalano

TRADEMARKS

The following literary material is provided for informational purposes intended to be used as a guide to a prior consultation with an attorney familiar with your specific legal situation. The author is not engaged in rendering legal advice, and proprietary rights discussed are not a substitute for the advice of an attorney. You should always seek the services of an attorney before attempting a computerization project as discussed in this manuscript. No attempt is made to designate trademarks, services, or related terms in which proprietary rights may exist. The initial capitalization for such a venture is significant and should be carefully planned before implementation. Inclusion, exclusion, or definition of a word or term is not intended to affect, or to express judgments upon, the validity of legal status, or to express judgments upon other possible and valid business solutions, or proprietary rights claimed for a specific word or term.

To order additional copies of this book, contact:
Xlibris Corporation
1-888-795-4274
www.Xlibris.com
Orders@Xlibris.com
37114

To the employees and contractors of Studio K Productions LLC, SonicImpact Records, and Catalano's Agency: For their drive, professionalism, and dedication to excellence

Acknowledgments

The following thanks are dedicated to the organizations and professionals who participated in the development and editing focused to increase sales and profits, hardware and software, and customized networking procedures and proven IT modifications.

- Microsoft Partner Program (MAPS) who supplied the software and knowledge base support for the study
- My company's clients who have provided me with the ability to continually grow and learn my technical craft
- My agency's employees, managers, and IT staff who have developed a highly respected organization and special workplace
- My professional associates who offered mentoring and valued guidance and counsel
- My professional family, especially Maria and brother Marc, who have always been my driving force to success offering reliable direction, with their ongoing business support and love.

Disclaimer

The purchase of computer software or hardware is an important and costly venture. The author and publisher have made every effort to insure accuracy and timeliness of the information contained herein. The author and publisher assume no liability with respect to loss or damage caused by the reliance on contained information and disclaims any warranties, expressed or implied. This book is not intended to replace a products documentation or capabilities of products discussed within the material. The specifications of computerized equipment and related software are subject to frequent upgrades modifications. All configurations of computer hardware and software should be discussed with the manufacturers prior to choosing any computer applications.

AUTOMATED NETWORK TECHNOLOGY:

THE CHANGING BOUNDARIES OF EXPERT SYSTEMS

A Dissertation
Presented to the
Faculty of the
School of Business Administration
Kennedy-Western University

In Partial Fulfillment of the Requirements for the Degree of
Doctor of Philosophy in
Management Information Systems

by
Carl P. Catalano
Miramar, Florida

Abstract of Dissertation

AUTOMATED NETWORK TECHNOLOGY:

THE CHANGING BOUNDARIES OF EXPERT SYSTEMS

by
Dr. Carl P. Catalano
Kennedy-Western University
Warren National University

THE PROBLEM

The main problem of this research approach is to identify appropriate practices to be used as a guideline for reengineering and implementing automated WAN measures into an operational multiregional business operation, while improving security precautions, applying a feasible Virtual Private Network (VPN) application hardware and software development strategy.

A secondary problem of the research is to increase business potential by determining improved client channel service and enhancing security intelligence beyond the traditional business expectations. Application software choices are even more difficult as a growing concern emerges towards the variability of information, which serves an important role in implementing security procedures. Application software generic perspectives extend important roles of WAN technology development guidelines.

METHOD

The overview of WAN security information experiments with the adverse effects of implementing automated technology to gain improved business advantage of general hardware and software solution applications.

The study also experiments with different alternatives to enhance valuable security protection when implementing intelligent machines. Resource motivation strategy is an important consideration for machine training review.

Effective automated business solutions coexists with an increased mandate and interest in faster and more reliable computer technology to satisfy the prerequisites of the organization and to determine ad hoc resource training requirements. To resolve these issues, the study focused on system design experts who have worked in the discussed areas and who provided insight into projections of the hypothesis. Test equipment, software tools, and analysis help establish a direction for changing obligations.

The review of literature polls several authors who contribute to a better understanding of automated technology and its implications on the workforce. The database focal point analysis is about technology experts who have significantly contributed to the field of study. The perceptions, insights, and judgments of writers contribute an understanding of the information and facts of the VPN application development. The literature database is focused on projections of related subject matter to enhance the execution process. Research results are compiled from a three-year reengineering project, which enhanced the IBM UNIX automated security and multimedia production system. In addition, the research method importance is to augment and identify appropriate practices to consider as a model for implementing VPN technology and to measure effectiveness of application software development.

External outsourcing and resource channel management add dimensional perspective and deliver increased production output potential. Automated network hardware and software tools are designed to monitor system activities. The guidelines discussed identify appropriate discipline, which comply with legal mandates and actively analyze automated system activities. Distribution of the WAN environment is a sensible direction that shapes the rapid-access exchange collective.

FINDINGS

Based upon the written works of Edward Amoroso, Elias Awad, Carolyn Begg, Harry Collins, Douglas Comer, Thomas Connolly, Jan Harrington, Vincent Jones, Peter Keen, Martin Kusch, Mark McDonald, Barbara McNurlin, Tere Parnell, Ralph Sprague, Alan Weis, and others, the review determined use of automated information systems have impacts on productivity, shift job responsibilities, redistribution of authority and power, while changing personal, career, and business associations. Technology implementation will influence the organizational structure and the channels of communications, which depend on applications used, and the specific problems they are attempting to resolve. The philosophy of the organization towards implementing WAN intelligence affects the job responsibilities and demonstrates how well system designers construct their system. Knowledge engineering can be automated but will certainly revise and affect the organizational resources. Objectives can be thoroughly measured after enough applications are integrated.

A sensible machine-training objective coexists with irrational design application expenditures. Effective ad hoc business solutions motivate a strategic discipline of automated network channel management. WAN administrative priorities establish heuristic business channels that influence the interfaced layer objects.

A paradox of security regulations plagues 21st century system architecture. Terminated and disenfranchised employees motivate a need for network administrators to determine an appropriate balance of strategy, consequence, employee consideration, and financial stability. We must approach automated machine management from a human perspective and branch into national alliance contention. VPN expansion will also increase corporate risk exposures as well as introduce novel ambiguities into labor laws.

Administrative Outsourcing Engineering

Contract Workers—Do they help or hinder a project; or just take time to get on track?

How can we work with them?

Case Study Example: A computer programmer is laid off when Studio-K Enterprises downsizes. We agree to pay a contractor to replace the programmer a flat amount to complete a one-time project to create a certain product. It is not clear at this time how long it will take to complete the project, and the independent contractor is not guaranteed any minimum payment for the hours spent on the program development. In addition, the project manager provides this contractor no detailed instructions beyond the scope of specifications for the product. The independent contractor who has now decided a written contract is necessary, which provides that the contractor is considered an independent contractor, and required to pay Federal and state taxes, and cannot receive employee benefits, files a Form 1099-MISC. Our contractor does the required work on a new high-end computer system that in-house staff budgets at $10,000. The contractor now works at home and is not expected or allowed to attend corporate meetings because the software development group has not cleared this potential security and intellectual property threat. Furthermore, the CFO feels that additional costs are necessary as a result of our already implemented Security Policy (http://www.techies.com). Recently a Minnesota-based company reported on some new research data on the new trend in tech employment. Apparently this Techies.com research project has found that employers are hedging their bets on the new economy by hiring contract tech professionals instead of permanent staff. It seems that the tech job market is looking a little more favorable these days. In an economic upturn, contractors are the first ones employers hire. Research has uncovered a trend in the way a tech worker is now hired. The economic situation of today is such where contract work is a leading economic indicator. When recovery comes, a lot of companies feel more confident hiring temporary contract workers to get through their staffing. On the other hand, when the economy spirals downward as many investment indicators recently display, as interest and resource costs increase, contract workers are usually the first group of people they let go. So a corporation gains a lot of flexibility; it appears that this concept is also cost-effective. The economic factor is certainly looking to come back although confidence levels aren't so optimistic; so corporate management is looking to contract workers to help them with their projects.

Tech workers perspective seems to be a less secure employment situation. The question becomes the reality of how should the technology worker view this situation from a long-term standpoint. The tech worker usually views this situation as a great opportunity and in a positive way of finding permanent work. Talking to people and building relationships seems to offer a project that is going to last 30-60-90 days and it's a project that you're qualified for; this newfound prospect may even challenge your sense to take the offered project. The opportunity of staying with that company is certainly increased by the fact that you worked with them already. The advantages for employers are trending toward contract positions because they only pay for the hours they need and use. You may need somebody thirty hours this week and ten hours next week, so it buys needed flexibility. You don't want to hire somebody full-time because you may only have a project that will last sixty days and you don't want to make a long-term commitment. The other advantage is long-term cost management, which is a long-term perspective that is often overlooked. When you try before you buy, it works for both the contractor and the employer. When you hire somebody that has worked for you as a contractor, his loyalty is already built up. You've already gone through the social process and you see an opportunity to have a strong relationship. Contract workers should position themselves to take advantage of contract work by networking and also using online web tools to post resumes to promote interests in contract work. You're responsible to take the job prospective search into your own hands. Professionals cannot wait around for the phone to ring. Employers do a good job of finding staff for positions, and professionals can't put their entire future in just one potential project.

As an independent contractor, you should cover your overhead. If you recently worked for $30 an hour and the next contract is just $25 per hour, you certainly want to cover every possible cost. Most companies recognize and compete for your skills because staffing company's mark up the labor costs to cover their administrative fees. Research has proven that a contract staffing company frequently will pay at least 1.5 times their direct labor budget as long as you are fairly priced for your skills. Techies hang out with people who have the ability to search a database via skill sets. On the posting side, we match the client's needs with the required skill level. Whenever there is a match to a job in their disciplines of skills, within their regional boundaries, our technology will push that position with predefined priority rules.

BENEFITS OF USING AN INDEPENDENT CONTRACTOR

When a business hires a worker as an independent contractor, it is often because the business intends to avoid the responsibilities attached to an employment relationship, most notably, workers compensation insurance, employment tax, and wage withholding responsibilities. The IRS examines Behavioral Control, Financial Control, and Type of Relationship. Facts that provide evidence of the degree of control and independence fall into the following distributions:

Behavioral Control—Instructions the business gives the worker; training the business gives worker. Facts that show whether the business has a right to direct and control how the worker does the task for which the worker is hired include:

- The type and degree of instructions the business gives the worker. An employee is generally subject to the business' instructions about when, where, and how to work. Even if no instructions are given, sufficient behavioral control may exist if the employer has the right to control how the work results are achieved.
- Training the business gives the worker. An employee may be trained to perform services in a particular manner. Independent contractors ordinarily use their own methods.

Financial Control—How the business pays the worker; the extent worker incurs business expenses which are not reimbursed. Facts that show whether the business has a right to control the business aspects of the worker's job. Independent contractors are more likely to have nonreimbursable expenses than employees. How does the business pay the worker? An independent contractor is usually paid for the job. An independent contractor can make a profit or loss. Facts that show whether the business has a right to control the business aspects of the worker's job include:

- The extent to which the worker incurs business expenses which are not reimbursed. Independent contractors are more likely to have nonreimbursable expenses than employees. Fixed ongoing costs that are incurred regardless of whether work is currently being performed are especially important. However, employees may also incur nonreimbursable expenses in connection with the services they perform for their business. An independent contractor often has a significant investment in the facilities he or she uses in performing services for someone else. However, a significant investment is not required.
- The extent to which the worker makes services available to the relevant market.
- How the business pays the worker. An employee is generally paid by the hour, week, or month. An independent contractor is usually paid for the job. However, it is common in some professions, such as law, to pay independent contractors hourly.
- Written contracts describing the relationship the parties intended to create—whether the business provides the worker with employee-type benefits, such as insurance, a pension plan, vacation pay, or sick pay; the permanency of the relationship of a worker with expectations that the relationship will continue indefinitely, rather than for a specific project or defined period.

Type of Relationship—Whether the business provides the worker with employee-type benefits, such as insurance, a pension plan, vacation pay, or sick pay. Set the duration of the independent contractor's services to the Corporation. Establish the details of the independent contractor's relationship to the Corporation. Prevent the independent contractor from impairing the reputation or goodwill of the Corporation. Whether the business provides the worker with employee-type benefits, such as insurance, a pension plan, vacation pay, or sick pay. When a business hires a worker as an independent contractor, it is often because the business intends to avoid the responsibilities attached to an employment relationship, most notably, workers compensation insurance, employment tax and wage withholding responsibilities.

Additionally, when a worker is treated as an employee, the employer may also incur costs related to pension plans, health insurance and other fringe benefits, workers' compensation, and state rules relating to employment taxes and regulation of the workplace, etc. A worker may choose to be an independent contractor because he wants to work autonomously, because he places low value on fringe benefits (e.g., health insurance), because that is the only way the business will hire him, or for other reasons. From a tax perspective, a worker may prefer being classified as an independent contractor as opposed to an employee because employee business expenses are deductible only as itemized deductions and are subject to a 2 percent floor of adjusted gross income. In contrast, an independent contractor may be able to deduct expenses as ordinary and necessary business expenses on Schedule C (Form 1040). When a business hires a worker as an independent contractor, it is often because the business intends to avoid the responsibilities that attach to an employment relationship; most notably, workers compensation insurance, employment tax and wage withholding responsibilities.

From a tax perspective, a worker may prefer being classified as an independent contractor as opposed to an employee because employee business expenses are deductible only as itemized deductions and are subject to a 2 percent floor of adjusted gross income. In contrast, an independent contractor may be able to deduct expenses as ordinary and necessary business expenses on Schedule C (Form 1040). Also, the deductions and/or benefits available to the independent contractor under a self-employed pension (Keogh) plan might be preferable than the benefits that would be available under the pension plan (if any) of the company receiving the services.

In order for a corporation to use independent contractors, they must meet the criteria used by the IRS to determine whether an independent contractor is, in fact, an employee. There are three main areas and eleven factors the IRS analyzes in order to determine the difference between an employee and an independent contractor.

Common-law rules. To determine whether an individual is an employee or an independent contractor under the common law, the relationship of the worker and the business must be examined. The extent to which services performed by the worker is a key aspect of the regular business of the company. If a worker provides services that are a key aspect of your regular business activity, it is more likely that you will have the right to direct and control his or her activities. For example, if a law firm hires an attorney, it is likely that it will present the attorney's work as its own and would have the right to control or direct that work. This would indicate an employer-employee relationship.

New tools/test capabilities and new work methods—what are these tools, test capabilities, and work methods that can improve your efficiency?

Software test automation can dramatically reduce costs and speed up time-to-market, although without the right strategy, expensive test automation tools may increase costs and delay progress of the project. The key benefits and tangible results will vary within the industries and certainly can become difficult to measure; the key benefit of software test automation effectively examines and can analyze the key milestones of the projects failure or create a model for success. Automation strategy helps to enable the organizations pitfalls and focus development with real-time ad hoc solutions to avoid potential problems. Action-based testing features the latest improvements to keyword-based strategy.

The achieved benefits are the reasons why test automation projects fail to achieve their objects. Automated tests take less time to fully execute than manual tests and can generally run unattended. A tester simply starts the test and increases test coverage on each testing cycle: Automated tests enable testing teams to enable teams to test more features in each cycle scope, and then test end users; test automation will never entirely replace the need for human testers because subtle system bugs are almost never detected by test automation, particularly usability faults. Ad hoc testing techniques, automated mundane tests, enable knowledge engineers to focus on using their creativity in more user-friendly classifications.

Despite the clear benefits of test automation, many organizations are not able to build effective test automation models; Chronicles test automation becomes a costly effort. There are a number of reasons why test automation efforts are unproductive pitfalls. Test automation projects fail to achieve their designed potential because of:

• Poor quality of tests being automated for model design
• Lack of good test automation framework and business processes

- Inability to adapt to changes in the System under Test (SUT) under significant change.

Automation framework is designed to record system history for developers to consider how to handle system variation; teams often find that the majority of their test scripts need maintenance. Stakeholders begin to lose faith in the results of the test automation. Poorly perceived value in the test automation process will certainly result in a corporate decision to scrap the existing test. Many teams acquire a test automation tool and begin automating as many test cases possible. Consideration to manage the test scripts and results creates reusable functions, separating data from tests and other key issues that allow a test effort to grow successfully.

Software test automation has evolved through several generations of tools and computerized tests to be run unattended, greatly increasing productivity. Engineers frequently require that an automated test be recorded again manually to alleviate issues with the capture-playback tools, as changes to the software require more complex transformations. Data-driven testing is often considered separately as an important knowledge base development tool in testing procedures. Keyword-based testing breaks work down even further by reducing the cost and time of a test design and execution, allowing all members of a team to focus on executable test automation using "keywords" that represent actions recognizable to end users. When the software under test is changed, revisions to the test and to the automation scripts are necessary as entire test scripts can be swapped into other modifications in the library as needed.

Automated testing templates and business rules based on the tested design provides the interface that enables nontechnical users to create tests by specifying low-level actions, and low-level keywords. Nontechnical test engineers and business analysts can then define their tests in a series of automated keywords, concentrating on testing rather than the scripting language.

It certainly is important to use high-level actions whenever possible. However, high-level actions are built on automation failure when creating a high-level action required by the modification. The organization should develop test standards that can be reused in the next phase faze of the process isensuring that the test process can be both the test process model and the end user product, resulting in higher quality more effective test procedure. Full Involvement of the testing team in test automation team members who are fulfilling the role of test automation engineers are often experts in testing fundamentals, the software under test, or the business domain; allows collaboration of both types of team members to contribute to the testing that is more effective.

Testers define tests as a series of reusable high-level actions, while the automation engineers focus on the technical challenge of implementing the system rules into user actions. Significant reduction of test automation maintenance builds a significant automation suite. Automation teams spend more time maintaining their existing tests than actually creating new tests. Testing significantly reduces the maintenance burden by allowing users to define their tests at the business process level; rather than defining tests as a series of interactions with the UI, test designers can define tests as a series of business actions. Processes will still remain the same so the test designer does not need to update the test. Improving the quality of automated tests follow systems architecture in a top-down approach that helps to ensure common methodologies:

- Different quality attributes being tested in business processes for consistency; once the test modules have been identified, define test requirements.
- Test requirements are critical because they force developers to address incident in the module, and each test case is associated to one or more knowledge base requirements; each test requirement should be addressed by one or more cases.
- Test developers can be concise in their test creation, explicitly defining the requirements so designers can start implementing results using either predefined actions or by defining new actions.
- Test designers should define their construct as high-level business processes, which allow the tests to be more readable than defined using low-level interface interactions.
- Many teams dive into test automation without first considering how they should manage the database libraries.

The challenge of sharing managed information database libraries is multiplied by a proven framework for organizing archives with a clear structure, preventing disruptions that can be caused by enabling remote sharing of database repository modules; actions that offer the full power of organizational testing including business analysts, technical testers, expert automation. Engineers and managers can provide a set of customized tools that will provide a knowledge base solution for design, automation management, test repositories, user administration, and valued action-based applications of effective, rapid response software testing methodologies that effectively integrates the latest software test automation technologies and offshore testing strategies into a comprehensive elucidation that fully capitalizes on the speed and cost-effective advantages of best practices in automation and integrated outsourcing.

Action-based testing improves test organization for better, faster, and more cost-effective software testing. Testing system architecture solves the biggest problems of system development although presents a significant challenge of producing and maintaining vast libraries of manual and automated test scripts. Development teams need a more powerful, more comprehensive approach to analyze and ensure application security in development—one that leads them straight from a high-level view of the problem to the source code itself.

Greater complexity—products and projects keep getting more complex. How do you meet the challenge? Is there any way to simplify? What can be done to simplify projects?

Industry statistics show that a large percentage of software security breaches happen at the application level, rather than the network or system level. These places increased pressure on IT staff to address security vulnerabilities during the development process. But without the right approach and tools, the job of identifying and fixing potential security problems in development can drag out the process way past critical deadlines—and still leave far too many security holes.

Developers can use an approach that automates attack simulations and identifies security issues during runtime. In short, developers need automated security analysis. Organizations must overcome a number of challenges that preclude inconsistent business processes that can fragment customer data and disconnect communication channels; inefficient or not properly trained staff lack the competencies and education required to deliver an outstanding customer experience.

The effectively managed systems can help develop the best customer experience that becomes a viable comprehensive plan to facilitate organizations to optimize their human resources in the direction of a valuable potential of processed information technology built around the customer, resulting in improved customer relationships.

Emulating—The secure channel network of virtual experience is advocating superior business results, with a contingency factor that is necessary for long-term growth and a competitive advantage. The customer experience is then a comprehensive prescription for achieving the organization's long-term goals for success. Superfluous outmoded modeling, networking, achieving certifications, discipline, and virtual educations allow for a growing globalized virtual community of valid e-commerce communication that helps supports Moore's Law of changing delays in support of user methodologies with third party developers, although it has been significantly improving. When it comes to new technology, customer support, networking, useful business solutions, and customer support—with quality that will inspire a strong partnership throughout the industry—we have discovered a valued discipline of tools for today's IT professionals. Microsoft is certainly my choice for developing long-term business relationships. Microsoft is certainly growing a globalized virtual community of valued e-commerce communication—allows for a growing globalized virtual community of valid e-commerce communication that helps supports Moore's Law of acculturated change.

Contents

List of Figures

List of Tables

Abbreviations

ADSL	Asymmetric Digital Subscriber Line
AI	Artificial Intelligence
ANSI	American National Standards Institute
ARPNET	Advanced Research Projects Agency Network
ASCII	American Standard Code Information Interchange
ASIS	Application Specific Integrated System
ATA	Advanced Technology Attachment
ATM	Automatic Transfer Mode
BIOS	Basic Input Output System
CAD	Computer Aided Design
CATV	Cable TV
CCITT	Committee for International Telephone & Telegraph
CEO	Chief Executive Officer
CLI	Command Line Instructions
COBRA	Common Object Request Broker Architecture
DBMS	Database Management System
DCE	Distributed Computer Environment
DEC	Digital Equipment Corporation
DHCP	Dynamic Host Configuration Protocol
DMT	Discrete Multi Tone
DNA	Digital Network Architecture
DNS	Domain Name System
DOS	Disk Operating System
DS	Digital Signal
DSP	Digital Signal Processing
DSU / CSU	Data Service Unit / Channel Service Unit
EEOC	Equal Employer Occupation Commission
EIDE	Enhance Integrated Drive Electronics
FCC	Federal Communications Commission
FDLE	Florida Department of Law Enforcement
FTP	File Transfer Protocol

GEO	Geostationary Earth Orbit
GUI	Graphical User Interface
HA	High Availability
HFC	Hybrid Fiber Coaxial
HSM	High Storage Management
HTML	Hyper Text Markup Language
IA	Implementation Agreement
IDL	Interface Definition Language
IP	Internet Protocol
IS	Information Science
ISDN	Integrated Service Digital Network
ISA	Industry Standard Architecture
ISO	International Standardization Organization
ISP	Internet Service Provider
IT	Information Technology
ITU	International Telecommunications Union
LAN	Local Area Network
LAP-D	Link Access Protocol-D
LU	Logic Unit
MMC MIDI	Machine Control
MODEM	Contraction of Modulator/Demodulator
NIC	Network Interface Card
NII	National Information Infrastructure
NIST	National Institute of Standards & Technology
NSF	National Science Foundation
OLE	Object Linking Embedding
OMG	Object Management Group
OSF	Open Software Foundation
OSI	Open Systems Interconnection
OS	Operating System
POST	Power On Self-Test
POTS	Plain Old Telephone Service
PRI	Primary Rate Interface
RAM	Random-Access Memory
RAS	Remote Access Service
RF	Radio Frequency
ROM	Read-Only Memory
RPC	Remote Procedure Call
SCSI	Small Computer System Interface

SMPTE	Society of Motion Picture and Television Engineers
SNA	Systems Network Architecture
SQL	Structure Query Language
TCP/IP	Transmission Control Protocol / Internet Protocol
TQM	Total Quality Management
UTP	Unshielded Twisted Pair
VGA	Video Graphics Array
VPN	Virtual Private Network
W3C	World Wide Web Consortium
WAN	Wide Area Network
WEP	Wireless Encrypting Protocol
XDR	External Data Representation

Chapter 1

Statement of the Problem

The current problems are WAN standards and the necessary tools to assist network administrators. The project reviews LAN system design for reorganizing business channels and improving automated customer services by implementing automated network intelligence. The three-year plan proposed is intended to develop a marketing and training strategy using a reliable UNIX / Windows 2000 XP client-server architecture of object-defined information. Exchanging distributed WAN technology with new government regulations increases the challenge of inventing security precautions to protect corporate exposure and adding advantage to customer service in a national market.

The second part of the hypothesis is to validate engineering skills and consideration of all viable network solutions including administration curriculum knowledge; and to resolve issues for information technology distribution, a feasible test for system administrators. The questions for review evaluates information protection theories of reengineering solutions by expert system designers and knowledge engineers who explained and illustrated the most important aspects of automated WAN technology available to system designers and help reveal a viable UNIX / Windows 2000 / XP software VPN application.

Administrative Priority

A review of firewall-gateway determined fiber-optic intelligence in broadband technology requires auto-sensing high-speed DSL to monitor traffic, channel footprint tracking, to manage accounts, and distribute resource production. The user security functions and operational characteristics are linked to quality client application, which are reviewed, restructured, and then analyzed for future resource objectives.

The knowledge engineers determined two important development priorities.

1. Review reliable, outdated, UNIX ANSI C source code in order to revise and improve machine security protocol to meet the challenge of business demands implementing system control mechanisms for high-speed WAN client services.
2. Evaluate automated WAN solution through in-house training objectives for trusted but unskilled personnel to communicate with regional leveraging from distributed information or review multiple VPN revenue generation opportunities—adverse affects of automated services, client retainer tracking, project management control, and internet publishing technologies.

Enterprise computer network proposals are at the core of the routine business activities. LAN terminology and their resource connection elements are important for reliable communication and employee education, which present exciting possibilities for the VPN enterprise. The very newest and fastest technologies require sophisticated DSL connections extending boundaries of computer science applications to the leading edge. Knowledge responsibility is finding factual austerity of information on any subject. You may be presented with a fusion of confusing, incomplete, self-serving, and even misleading information on a variety of WAN systems. Clear, concise information analysis of the science defines general WAN type, minimum performance, and a variety of leading design schemes.

Overview of the Study

The research-discovered value in hardware, software, and sensible production solutions regulates a database of over three thousand licensed professionals collecting employment consulting and WAN training alternatives. The Florida Department of Law Enforcement (FDLE) approves applicants, and Florida Chapter 400, Part IV, regulates the industry.

Employment impact consequences or poorly reread management considerations initiate fear that will continue as long as corporations have insecure managers or improperly introduced technology into the organization, challenging the reasonable and potential operation strategy of information system design. Corporations are now recognizing knowledge engineering as a valuable asset, which adopts the future perspective in policy management.

To understand the problem, the researcher's evaluation examined a UNIX based implementation project, which reorganized the ANSI C 3B2 system architecture and developed improved firewall security precautions.

Network scalability fused with appraised resource management guides a sensible hypothesis review. Network administrators define attitude through system control regulation, tracking employer's responsibility to define, regulate, and monitor the subject matter of an intelligent database. Control matrix supplements observe network activities and thereby comply with expanding government regulations.

Human resource potential measures the network administrator's ability to shape the environment, stimulating human resource capability. The standard query theory determines moment-to-moment assignment allocation criteria to minimize service time under cost constraint and suggests that human communication misunderstandings degrade efficiency more than limited processor or bandwidth speed. The study determined that new areas of security concerns challenge the network administrator to understand WAN technologies available, evaluate considerable expense, and motivate human resource development, which has evolved into integrity protection of the organizations important infrastructure.

Automated system intelligence adapts to programmed priorities, adjusts diagnostics, and offers intelligent suggestions to correct system malfunction while keeping system protocol within fixed tolerance limits. Initial attention is a non-synchronous critical time constraint remodeling into a rapid-access synchronized WAN solution.

Infrastructure protection application and reengineering design logic are presented to describe the common network diagrams and WAN IP channel technology. Utility functions task client / server application software and system operations are analyzed. The overview of network technology demonstrates improvement in customer service functions compared to average LAN security system solutions or management training objectives. A traditional star network design was expended as illustrated with important firewall protection consideration over prior UNIX security protection.

The first part of the hypothesis will evaluate the standard network system architecture then apply appropriate logic feasibility for considering WAN security alternatives. The goal is to execute, analyze, and contrast resource perspectives suitable for VPN designers.

IP Significance of the Study

There are numerous combinations of an application software and hardware available to system managers today. Automated technology emerged the industrial society to the IP information-based channel infrastructure.

Widespread implementation (1990) of the Dynamic Host Control Protocol (DHCP) simplified the host network, although it increased the complexity of IP security. A Bootstrap Protocol (BootP) device is without permanent storage, printers, X terminals and routers, IP address and start-up configuration parameter or hardware, and corresponding IP addresses. The DHCP server assigns the IP address and configuration parameters to network devices, although with a significant difference:

The intelligent DHCP server is not aware of the device hardware address. Instead, the DHCP pools or groups of IP addresses are dynamically assigned to clients on a first-come, first-served basis. Specific IP protocols are fundamental properties of information security integrity, will expose or alleviate serious damage of vicarious liability and harm to the organization's fiscal stability.

Statement of Objective

The primary objective is evaluating resource management policy by determining the best alternatives for WAN management. Administration is interested in electronic commerce advantage measuring past, present, and future technology benchmarks, and motivating a sensible user training solution.

Companies depend on the reliable information infrastructure to expand traditional boundaries of conducting business. System dependencies articulate software application as an important role in intelligent security. Information management likewise has become a significant concern to network administrators.

The objective is divided into two intangible goals:

1. Determine the sensible system model to justify corporate liability exposure of increased business operation expenditure and to stimulate future growth incentive.
2. Review strategic measures to evaluate resource training objective and protect the network security implications of venture in order to determine the correct administrative policy and business decisions.

Rationale of Study

A bug found on most twentieth century machine code is reinventing the system model, the type "S" corporation, expanding it into a national client Web site-hosting plan. A frame relay firewall monitors a DHCP gateway of secure access Dynamic Network Server (DNS) consulting or multimedia production. Satellite Digital Subscriber Line (DSL) locations are important in channel linkage implementation to increase regional market share. Key issues are infrastructure confidentiality properties and e-commerce security elements.

The study examines previous automation solutions to determine the best employee educational alternatives for WAN distribution or discover possible adverse implementation affects. A politically active administration retains lobbyists for national associations and market sensitive alliance networking. Human resource participation is the rationale for improved channel leveraging.

The first part of the hypothesis is focused on the question of feasibility for the proposed software and hardware alternatives and will illustrate information management perspectives derived from expert's opinions in the industry. The study will illustrate important aspects of WAN information distribution and experiment with Mackie (3.0) automated production implemented in a UNIX / Windows based VPN reenergized strategy. The study explores automated customer service feedback control of resources measured through task monitor or corporate fiscal growth statistics.

To better illustrate the e-commerce historical perspective challenge: "Companies will go outside the private network and telecommunicate, while on the road, determining alternatives, to secure a profitable market share alternative, objectively measuring effective human resource motivation solutions." (Wolf, Scott, Erwin, 1999)

Rewards for choosing the correct alternatives prove long-term goal success factors and increase the magnitude of network value. In the short term, a reliable UNIX-Microsoft, client-to-server, peer-to-peer star network is deployed. System administrators rely on a decision network of multichannel, server-routing application software. The long-term functional solutions forecast tracking of accounts or staff scheduling, isolating key factors of issues that necessitate an immediate resolution. WAN design security functions are an important aspect of the systems logic infrastructure. The strengths and weaknesses of the automated network intelligence will also be analyzed. Outline includes UNIX firewall security measures, management issues, and evaluated alternatives in system design

Scope of the Study

Concrete judgments achieve expected resource outcomes when channel implementation is measured by employee progress towards specific goals of motivation. Quantifying channel standards are difficult to measure because each system is objectively defined by a particular problem, which is eventually resolved. For example, a consulting firm or accessible source publishes measurable objectives such as sales, growth, and profit and defines market share holdings annually. The gross margin of service channels contemplates issues for courts seeking to apply the appropriate standards.

Limitations of the research do exist because the researcher works daily on a UNIX / Windows 2000 servers and other reliable solutions were excluded and are considered a reliable approach in WAN distribution. The value of technology cannot be thoroughly measured and cannot supersede or predict every aspect of information vulnerability. The study does not imply this is the only suitable application for automated WAN development.

The review of literature approach to development of automated technology will stress the importance of information protection and human resource motivation. A prescribed financial model motivates automated system designers and the researcher is limited to available resources. Because of these facts, evaluating every possible WAN alternative is not feasible and it limits business approach perspective to assist each system administrator. A clear intelligent WAN understanding is developed to readers familiar with LAN information distribution and network security issues. Interesting related articles to reference are listed in the bibliography section.

Framework of the Study

Chapter 1: A statement of the problem, administrative priorities, overview of the study, IP significance of the study, objective, rationale, scope, with particular consideration of security issues of feasible VPN, outline of study with import LAN terminology are introduced and defined.

Chapter 2: The literature reviewed VPN experts who explored similar issues of World Wide Web Consortium (W3C) channel management development. Respected authors also explained problems associated with implementing VPN technology. The UNIX / Windows 2000 server environment represent recognized trade-offs in protection, management, heuristic properties, resource perspectives, observable facts, Internet perceptions, IP routing, standards, and sharing intelligence.

Chapter 3: The method of developing the hypotheses of the study is built upon the researcher's three-year VPN implementation experience. UNIX/Windows based software applications are reengineered for enhanced security precautions and model consideration. The approach techniques described relate to database, design considerations, interface definitions, environment security settings, and their applied applications.

The implemented network before and after reengineered characteristics included firewall security functions: seven-layer stack architecture, server and storage system relationships; and the basic interface function key macros provide a common VPN application exchange design.

Chapter 4: The data analysis provides a brief overview of the VPN reengineering experience including a detailed description of the security utilities and design component functions. The study discusses the firewall application software, system hardware, high availability requirements of enterprise consulting, software requirements, and implementation details of user security functions, TCP/IP issues, and many other unexpected dependencies implementing voice and data recognition internetworking with intelligent automated supervisory control. Cisco networks, Annex, and service routers are integrated into Internet applications. Electronic process shapes actions into a high availability and enterprise network design. The security assessment of the reengineering application is also covered.

Chapter 5: Provides a final summary of the analysis with supplementary recommendations and conclusions that are drawn from the WAN application design. The reengineering of firewall protection considerations and implementation results are discussed with the recommendations that may prove useful for further projects in resource management or VPN implementation.

Appendix: Figure 1 illustrates database of significant VPN technology covered in research. Figure 2 illustrates telecommunications changes for VPN services. Figure 3 is IBM's common terminal-to-mainframe network architecture. Figure 4 is the seven-layer Asynchronous Transfer Mode (ATM) model. Figure 5 illustrates the star network components. Figure 6 illustrates the original LAN reengineered framework. Figure 7 is the original cable-extension method. Figure 8 illustrates the new application software architecture. Figure 9 illustrates the new distributed WAN architecture. Figure 10 illustrates the enhanced management framework. Figure 11 illustrates the Data Service Unit and Channel Service Unit (DSU/CSU). Figure 12 is the new VPN architecture. Figure 13 illustrates the new multimedia production studio framework. Table 1 represents the review of literature benchmarks in database. Table 2 represents the survey of regional tax analysis. Table 3 is the basic command functions and key macros of the new management information system.

Bibliography: Provides a detailed description of the references contained within the research review.

Definition of Terms

Ad Hoc—On-the-spot solutions neither anticipated nor recurring.

Analog—Describes a device of continually discrete changing values by a continuously variable physical property such as a voltage in a circuit.

Architecture—The overall design and construction is part or all of a computer system including factors of processor size, hardware, and sequence of bytes.

Artificial Intelligence—An attempt to build machines that can make like humans. Techniques involved from the research to solve complex and problems through a symbolic reasoning problem solver.

Asynchronous Transfer Mode—A method used for transmitting voice, video, and data over high-speed LANs and WANs with continuous fixed length packets transmit data at 53 bytes, 5 to control function, 48 for data.

Automation—The automatically control operations of an apparatus, a process, or a system by mechanical or electromechanical device which mimics the human origin of observation, decision, and effort

COBRA—The Common Object Request Broker Architecture from the Object Management Group whose members distribute communication between objects into applications, regardless of the way that they were written and the hardware platform they will run.

Collaboration Software—A set of network definition-based applications that allow users to easily exchange and share information

Compilation—Translation instruction into machine language.

Database—The collection of related objects, tables, forms, reports, queries, in Scripps, created in organized by a database management system of practically any type of collection of facts and system rules.

Digital—A device that represents the values in the form of binary digit bits.

Digital Signal Processing—Chip is integrated into several cards, modems, and videoconferencing communications, image manipulation, and data acquisition application.

Encryption—The process of improving information security in an attempt to secure unauthorized access and during protects data transmissions. This process is a reversal of decryption.

E-Commerce—Commercial electronic value between an enterprise and external entity, an upstream supplier, alliance channel, or a downstream customer over a universal exchange medium.

Enterprise—The term used to encompass the entire business group or corporation, including local, remote, and satellite offices.

File Transfer Protocol—The TPC/IP Internet Protocol transferring single or multiple files from one computer system to another.

Firewall—A barrier established in hardware or in software, or sometimes in both, that monitors and controls the flow of traffic between the two networks, usually a LAN and the Internet.

Frame Relay—A standard for packet switching network protocol, at speeds of 2 Mbps and improved bandwidth provides better efficiency and higher throughput.

Gateway—A shared connection between mainframes and servers to link related systems or provide large packet switching networks, whose communication protocols are different using a combination of hardware and software tools operating independently with processor, memory protocols, and architectures to convert data.

Heuristics—Rules based on years of experience

Hierarchal Storage Management—A combination of several types of storage systems nourished by intelligence of organized information, which reflects priority in a problem domain.

Hybrid Network—A collection of technologies such as frame relay, leased lines, and X.25.

Informix Software—A major supplier of object-oriented and relational database products.

Infrastructure—The physical hardware used to connect computers and users, transmission media, which include cable television lines, satellites, and radio frequency antennas, routers, etc., that control transmission paths; also can send, receive, and manage the transmitted signals.

Integrated Services Digital Network—A standard for a worldwide digital communications network richly deserving to replace all current systems with a completely digital, synchronous, full duplex transmission computers; and devices connect through simple standardized data transmission interface. They can transmit voice, video, and data simultaneously.

Intelligence—Acquired and applied knowledge through reason.

Interface—A bidirectional communications between system and users.

Internet—Expanding public access or electronic media used for consumer transactions and business-to-business relationships.

Knowledge Engineer—A versatile professional with competency in technical skill system tools including current methodologies, interpersonal communications, and organization of specialist knowledge that represents the knowledge base.

Leverage Resources—The primary fuel in the growth stages of business. A prerequisite to building a legitimate, thriving, and visible growth-oriented professional access of system support services.

Link—A web page or hypertext document connection between one element and another in the same or a different document relationship or between semantic concepts, nodes, and descriptors.

Logic Unit—Protocols developed by IBM to control communication and in systems network architecture: LU type 0-4, type 6.1, type 6.2, and type 7.

Mainframe Computer—a large, fast, multiuse computer system, often utilizing multi processes, designed to manage large amounts of data in complex computing tasks

Network—A large group of computers and peripheral devices connected by a communications channel capable of sharing files and other resources among several users in a peer-to-peer LAN or WAN.

Object Linking and Embedding—Microsoft protocol for applications to exchange and communicate using data objects. Data representations can be embedded or linked. Separate copies of the data in the original embedded copy changes will not be altered in the compound document in less embedded object are updated. If the data is linked, only one company of the data exists. Changes made in the original document will be made automatically in the compound document

Proxy Server—a software package running on a server positioned between internal network and the intranet. The proxy server filters outgoing connections so that they appear to be coming from the same machine in an attempt to conceal the underlying internal network structure blocking intruders. A system administrator can also regulate the external sites through which users can connect.

Reengineering—Redesign or reimplementation of aging software to newer recognized technologies.

Resource—Any part of a computer system that can be used by a program as it runs. Resources may include memory, hard and floppy disks, networking components, and other operating system peripherals, as well as queues, security features, and other less well-defined structures.

Router—An intelligent connecting device sends packets to the correct LAN segment then to their destination. LAN segments reference OSI reference model for peer-to-peer communications. Routers can use similar or different networking protocols and can be central, peripheral, local, remote, or internal. Routing is the process of directing packets from a network source node to the destination node.

Shell—In UNIX, the processor commands are from the user, and the shell interprets, and then passes to operating system for execution. In recent years, several popular shells are developed for an inference engine knowledge base.

Standards—Specific, mandatory requirements to measure and address the policy in addition to lead and measures elements.

Synchronization—The timing of separate elements or events, which occur simultaneously. In computer-to-computer communications, the hardware and software must be synchronized for file transfers.

Telnet—The terminal emulation suite of protocols for a UNIX environment that provides remote connection services such as, DEC VT-52, VT100, and VT 220, etc. emulation terminals.

T1—A long distance, point-to-point circuit providing twenty-four channels of 64 Kbps with a total bandwidth of 1.544 Mbps. The standard frame is 93 bits long, made up of twenty-four 8-bit voice samples and one synchronization bit. Transmits 8000 frames per second and is available in 64 Kbps increments.

Three-tier Application—A software application of client components, which usually contain a graphical user interface, or server components that contain processes of form business actions, or the database components, which store the business data.

Topology—The network map illustrates a physical description of cable connections to workstations, nodes, routers, and gateways. Networks are usually configured in the bus, ring, star, or mesh topology. Local messages are transmitted to user on another network.

UNIX—Novell's Netware, conceived in 1969 at AT&T's Bell Labs by Ken Thompson and Dennis Ritchie. The multitasking operating system is the most widely used general-purpose operating system in the world.

Utility—Specific purpose applications useful in system domain.

Virtual Channel Connection—In asynchronous transfer mode a logical connection between two end stations can be either switched or dedicated.

Workgroup—Groups of individuals who exchange work together and share common files and databases over a local area network.

World Wide Web Consortium—A huge collection of hypertext pages on the Internet.

Chapter 2

Review of Related Literature

Introduction

The discovery within the last decade of three hundred years of technology systems is among the most important and exciting revelations in modern civilization. This chapter provides a review of related literature for the described problem and provides useful insight to the important properties and issues of LAN development, describes the average computer network, and will introduce WAN security vulnerability. The review measures the WAN design trade-offs explored in the venture and provides useful protection and resource control considerations of intelligent machine development. The focus of Internet security is an important design characteristic of managed scalability, often viewed by investors as critical for protecting the financial perspective model, although shareholders expect legal experts to clarify statements of security, so that institutions must have a protected security posture in order to protect shareholder value. (Giarraputo, 2001)

The research review continues and expands on the tantalizing mechanized inventions of Leibniz (1600), who sought a number system easier for a counting machine to handle than the decimal system. He hoped to find a way for a machine to convert the decimal system into a binary numeration process. Because of his inspired work, the basic channel requirement perspective was recognized.

The first recorded industrial application of automation was the fly ball governor, a device that regulated the speed of a steam engine. James Watt (1787), a Scottish engineer, constructed this device. His work inspired investigations into fluctuating dynamics of the real world engine application by effectively controlling a routine machine activity. Design strategies are based on the analytical engine punch card computing,

developed by Babbage (1830), which contained the basic elements of an
automatic computer-storage and working memory.

Given the dynamics of the emerging technologies, Hollerith (1924)
founded the Tabulating Machine Company or International Business
Machines Corporation (IBM). Fundamental building blocks are required
to preserve and protect data resource as an important aspect of business
operations. John V. Atanasoff (1939), an American mathematician
and physicist, constructed the first special-purpose electronic digital
computer. However, Howard Aiken, a Harvard University professor
(1944), expended on his ideas and built an improved digital computer,
the Mark 1, which was controlled chiefly by electromechanical relays (i.e.,
switching devices). Diebold's research (1950) improved system efficiency.
(Appendix, Figure 1)

Capable information systems today are based on general-purpose
applications cascading the components and available resources of the time.
The fundamental goals of our predecessors have now inspired designers
to develop information security distributions built on trusting a computer
system to preserve and protect its valuable resources beyond the traditional
scope requirements. Knowledgeable associates combine resources and
disciplines to improve adequate access control. The underlying principle
design of a secure WAN environment should be dependable and behave
as expected, which are administrative issues and resource concerns built
upon specialized disciplines of mechanized invention.

Self-regulating machines, electronic sensing, and computing
techniques were often modeled on human brain functions to address
intelligent system application requirements in authentication, access
control, and user authorization, which remain significant problems when
implementing equipment. Novel concerns and security requirements
inspire the need to understand information distribution technology.
Most system problems discovered are based on results of a structured
analysis that includes analyzing application software, system architecture,
and improving the network environment. Balancing today's financial,
confidentiality, data integrity, and rapid availability escalates rapidly with
increased network vulnerability.

Today's network computing expands on recognized technology
benchmarks based mainly on three fundamentals aspects: hardware,
software, and economics. Hardware developments are generally
predictable, but software availability and compatibility issues have a

significant effect on the adoption of technical advances, while economics affects its wide-scale deployment. The level and duration of system training is also a factor with dependence on the user's knowledge level of the system attributes. System requirements range from simple to advanced systems queries in high level embedded C++ programming. Artificial Intelligence (AI) may take several weeks to implement.

Moore's Law: For the past two decades, advances in computing have followed Moore's Law, the observation made by Gordon Moore, a cofounder of the Intel Corporation who stated that the power of microchips doubles every eighteen months or so. Moore did not actually say this, but it is a practiced rule of thumb.

Administrative Sources Description

1. Review reliable, outdated, UNIX ANSI C source code in order to revise and improve software security protocol to meet the challenge of business demands implementing system control mechanisms for high-speed WAN client services.
2. Determine through in-house training solutions for trusted but unskilled personnel to communicate with regional leveraging from distributed information or review multiple revenue generation opportunities—adverse affects of automated services, client retainer tracking, project management control, and Internet publishing technologies.

Information distribution integration and protection strategy is a dimensional discipline utilizing management and technical specialties, which may have impact on every aspect of modern society. Technology experts agree that measured management improves client service channels and in-house training development projections. Analysis of the industry determined Web site publishing, a key factor in alliance channels, motivates client relations and Internet marketing strategy although encourage a tendency for managers to use knowledge, as leverage, to resolve what they feel is the right solution. Technical personnel seldom understand the diverse nature of WAN clients in the southeastern United States, which are skilled professional providers in a highly competitive industry.

Management is interested in people solutions, policy making, and industry recommended guidelines. A VPN environment is expected to resolve a premium electronic information exchange service for the status-conscious. In addition, WAN customer convenience is distributed without intervention over large geographical regions. Resource control reviewed by engineers quantifies statistical analysis for system model information and furthermore improving channel development on a national scale.

Nevertheless, implementing automated network technology, even with such advancements, presents a negative aspect to employers. Since the 1970s, the age of the microchip, change has been drastic, and microeconomic trends are spiraling out of control. According to the editor-in-chief of *US News and World Report*, Technology has favored the educated and the skilled. Typical college graduates jump aboard a discipline without future job possibilities. Recent trends in computer knowledge affect society by reducing the need for unskilled labor. "The danger of the information age is, that while in the short run it may be cheaper to replace workers with technology, in the long run it is potentially self-destructive, because there will not be enough purchasing power to grow a thriving economy." (Zuckerman, 1995)

Global competition and the decline of unions across the nation—mean fewer unskilled jobs—invite government defense cuts and inspire international scrambles over wages. Furthermore, government regulation potential increase FTP transmission speed initiative and encryption cascade expenses for future Internet technology.

The correct application for software, tracking management, and system discipline policies measure automated system development. Software choices cause various sorts of anxiety in administrative decisions. Although skilled training and risk exposure requirements are greater, the benefits of improved information tracking provide an effective information distribution center.

The system architecture improvements discussed are valuable solution to these problems cultivating a message to employees willing to conquer the anticipated challenge. However, administrative business directives continually dispute the management needs and technologists' recommendation. One school of thought is the technocrat should expand their knowledge of management risks and designed infrastructures should adequately describe the administrative incentive: "Computation and communications together with a sensible technology is enabling a distributed network environment." (Mokhoff, 2000)

Computation and communications interfaced with the "sensing technology" is enabling a distributed network environment that uses the Internet as its backbone. Information read by sensors is fed directly back to the system, enabling the machine to adjust its operation as needed. Although client services are improved, training and hardware compatibility issues delay the project progress through increased costs, precluding resolution of the numerous system possibilities.

Security Considerations: Computer security threats and vulnerability assessment are the primary goals for policy modeling, countermeasures, or database protection. It is important to recognize enforcements, mechanisms, and models. A security policy can be defined as "a finite set of rules which delimit the accesses that can be made to objects. In addition, the implemented security policy defines the security requirements for a given system." (Amoroso, 1994)

Ad hoc security solutions frequently determine the network administrator's role. Confidence issues plague Internet security policy concerns that are activating a high-level of apprehension directed towards neural sensing technology. Knowledge engineers revise the managements set of directions and operation procedures. Knowledgeable individuals Internet presence popularity alters the marketing strategy of community standards and launches a successful agenda for products and services. Modern civilization communicates with up-to-date system architecture, collecting processes by combining numerous independent local networks interfaced to several manufacturers or intelligent machines. Clusters of data are built on a dial-up network frame accurate routing system.

Wireless transmission is another popular and convenient solution, although digital cloning presents expanded security concerns that plague data distribution methodology. As one technophile relates, "In wireless, we have created another gateway and somebody in that gateway has access to all the traffic. It does not matter if the information is encrypted in your cell phone. If someone gets into an Internet system, you might as well throw the security out. It's not the airwaves that are the problem, it's the gateway." (Giarraputo, 2001)

Hackers are the terrorist invaders of privacy and can add destructive viruses to severely damage a corporation, cause extreme financial deficiency, and confuse routine channel node activities. Protecting interests in enterprise File Transfer Protocol (FTP) is motivating government regulations, such as taxing for retail goods and levels of encryption, with increased security requirements. (Appendix, Figure 2)

The entire information revolution relies on relaying a merchant's information without unauthorized access. Information destruction and fraud are common crimes committed on the Internet. Accessible business associations impose greater security risks to the organization although data encryption utilities improve access methods. The rate which hackers invade corporate systems are increasing, however popular hacker deterrents such as systematic key encryption, mirrored passwords, encrypt and decrypt data channels as public-key encryption is generated, one key encrypts data, and the other decrypts. Each username and password allows client users to access network files or to enter a personal gateway into an Intranet Web site. Security breaches occur if skilled intruders are allowed to crack passwords. Intruders enter through the firewall gate and are capable of disrupting tracer surveillance, although sensor triggers and algorithms are powerful enough to follow a hacker's tangled itinerary back to its origin. (Rao, 2001)

Today's information infrastructure consists of enhanced network firewall security enforcing protocols of network access. The firewall gatekeeper processes data to different sensing mechanisms, authenticating traffic and restricting network intruders according to access control policy. The firewall administrator stores access permissions and is taught to know specific priority restrictions. A firewall's footprint tracing tool is an important profile, logging and providing editing functions, although firewall protection strategy only protects against intrusion directly through the frame relay gate, which translates into inside intruders on the wrong side of the gate. Network administrators agree that outsourcing is the best method of deploying firewall protection because security experts prevent firewall contrition and attacks on the corporate network to evade system hackers.

Feasible business solutions are designed for everyone's benefit and system improvements controls enforce administrative policy and procedures. Repetitive, simple jobs are less tedious without the hazards of job injury or insurance claims, thus avoiding increased employee expenses. Furthermore, WAN business tracking and operation expenditure analysis identify potential risks, which decrease the likelihood of exposure or threat of an unmanageable technology environment causing employee frustrations with decreased user privacy incentives. Knowledge of a security policy helps determine the balance between system cost, convenience, and new management policy requirements. The Windows shell environment provides the consistent interface to the organizational clusters of defined objectives, distributing appropriate design considerations and encouraging resource communication awareness of implementation policy Automated security intelligence goes beyond the capacity of computer users responding to routine intervention.

Resource Perspectives

Primary: The reinventing procedure depends on the nature of measured outcomes of a primary business function to reduce costs or general-purpose network training and to promote improved system performance detail. The initial corporate member database review determined a current table of contacts and qualification information, including address, phone numbers, etc., categorized in each geographical area. In addition, the expanded service plan provides a 24-hour knowledge base of client support assistance.

The financial consequences of assurance technology systems are startling and disruption of subsystems can cause increased vulnerability to the organization. Research into the issue determined technology perverse workers tend to design, build, or reengineer a scheme to their advantage, prevent user friction, or monitor abuse to systems security, as well as utility diagnostics. In addition, they enable lower sale price, thus increasing profits, although another perspective claims that many people expect an increase in automation to cause more unemployment. Technology experts have a problem in determining valid people solutions then are quick to jump to technical solutions, which resolve the issue. (Truxal, 2001)

Secondary: Automated service initiative has expanded the choices available in the vulnerable path of channel complexity for our computer system. A high level of expertise to deploy effective regional and national resource alliance channels is a crucial development issue and a sufficient distribution strategy of expanded automated customer service, thus supporting operation of a VPN and leveraging a share in future product results.

System is evaluated in terms of its effectiveness on people, procedures, and performance of the business endeavor. The main areas of concern are quality of decision making, attitude of end users, and costs of knowledge processing. User education, training, and participation in the building process can help reduce resistance to change. The expert systems interface requires extensive training to master various methods and advance a proficient structure.

Management Trade-off Issues

Computer users clamor for speed and for high-speed broadband access to gain advantage, in particular on demand, linked through a DSL or Hybrid Fiber Coaxial (HFC). Software and hardware examination of practical experience requires a database of contacts and implementation strategy within a defined WAN solution. Clients appreciate a professional and less risky method of telecommuting and leveraging in-house automated information tracking. Alliance proposals are an important ingredient of strategic thinking, although cooperative partnership alliances are part of a contemporary strategic methodology. Joint venture formation between U.S. companies and international partners has been growing by 27% annually since 1985. (Ernst, Bleeke, 1994)

Managing an automated workforce is changing at an accelerated rate due to unprecedented environmental complexity. The professional discipline is indifferent to conventional motivational strategy and determines ad hoc alternatives. The administrative directive of cultivating a diverse workforce requires appropriate cultural poise merging ethnic tension generated by employee workgroup sessions. Associates gather information from a variety of daily observations that motivate human resource evaluation through meetings, where conflict resolutions are practical.

According to some experts in the field, acceptance relates to the perceived impact of the new system or status in the organization. Resistance has much to do with an individual's personality, the organizational structure, and relations with the area where the expert system will be installed. Turmoil results from merger layoffs causing disenfranchised staff and ripples from turbulent leadership direction, generating an unstable environment condition and creating a workforce difficult to keep productive, focused, and harmonious. Employees perform daily in turmoil, cultivating diverse change in corporate relationships. System training direction is meant to allow the company to gain a strategic advantage if legal actions ensue. (Cascio, Zammuto, 1987)

Lawsuits alleging sexual or racial harassment generate hostility, as do age and handicap discrimination. The balance of laws, regulations, and corporate policies are the center of a conflicting and ambiguous automated industry. Risk is rare without a firm grasp of the problem. The network manager is tentative, establishing a performance expectation, implementing a procedure of control and performance evaluation, and a standard measurement of discipline.

Common interests in business are a sensible competition strategy. In addition, the survival of business encourages collaborative partners, allowing a small firm to extend into new markets without additional workers. Network-enabled organization solutions downsize payroll expenditure, increase growth, and form outreach divisions. The strategic cooperative autonomy is flexible enough to handle corporate growth without increasing expenses.

Implementation of an expert system is the initiation of a new order of things. Management and personal relationships differ from culture to culture and typically conflict. Resolving emotional conflict with top executives usually consumes vital resources between upper management, harming the organization. People become anxious when they do not know what the new system will do and how it will affect their current occupation or their future career plans. The result of this anxiety is stress and further resistance to change.

The possible solution to this dilemma is alliance training and development recognized as a major component of system implementation training. Configured partnership cooperation exchange does assist organizations to achieve intended goals in today's business. Furthermore, the enterprise information executive benefits from political lobbyists to gain insight about future connections. The long-term business goals are inspired by beneficial component links in the expanded alliance environment. Human resource potential generates enormous energy, nourishing a relationship to strengthen defining mechanisms to mimic agreements.

A successful corporate perspective addresses the long-term production plan. Short-term goal analysis seldom justifies maintenance expense or warrants a substantial financial investment. A typical rationale of partnerships is to gain enhanced knowledge, expand business, or motivate system competence. Merging cooperative projects compete for interests in the same market. Marketing and system design confuse the balance of autonomy with adapting to predecessors. Combining resources improves marketing potential critical to selling product. A company struggles for financial support and often operates behind competitors. Expanded geographical divisions alter corporate incentive or industry demand, suggesting immediate access to new channel connections.

In a large sample study, Harrigan (1988) examined the influence of partner asymmetries on joint venture success. Performance is measured based on joint venture duration, survival, and manager assessments. The study suggests alliances between similar firms tend to be more successful than asymmetric partnership. A comparison of joint ventures concluded that dominant influence in joint ventures is more successful than balanced partnership, demonstrating favorable performance. Distribution is relative to the size of the organization and the stability of the joint venture. The traditionally combined venture concentrates on unstable alliances, marketing difficulties and after-sales services. Theoretical analysis emphasized the importance of several strategic alliance distributions potential. The stability mechanism is a system value in the creation process.

Structuring joint flexibility is modeling proven outcomes of long-term environmental influence. Managers voice opinions, forming a coalition of product, distribution scope, breadth, and purpose. The alliance industry structure has a significant influence on workforce performance. Joint strategic factors imply strategy and structure of execution rather than isolating organizations' earlier recognition of the influence of modern civilization. Commercial appreciation in a fluid business community is an uncertain process. Sorting numerous possibilities determines a future business course, decreasing the likelihood of project failure. Longevity of relationships acknowledges a task taken up together. Profitable stability creates newfound trust, winning in a competitive marketplace, and meeting and confronting differences.

Executive data streams fuel interest in a successful alliance position, dramatically altering business growth potential. The technocratic generation of hardware and software is causing painful uncertainties for consumers. The rapid acceleration of technology development clearly articulates the rules extending sensible margins of WAN environment. Agreeing to details of future perceptions stimulates growth between strategic partners. Common alliances improve the tandem operations of commerce design.

Combining organizations will systematically increase human resource performance. Damage from a complex managerial dispute is reduced if associates learn together through a sensible solution that is flexible enough to change direction. Furthermore, partnership alliances promote products and services in unfavorable Internet conditions and unify aims for beneficial coexistence. Joint ventures can improve cash flow if partners unite, pool resources, and expand into a strategic coalition. The coexistence agreement of defined negotiation is a trivial although genuine automated WAN design philosophy.

Observable Facts of Channel Distribution

The first war of the twenty-first century was about information gathering and smart weapons. A specially designed electronic information system provided automated guidance to precision missiles, although snipers walked the desert with simple rifles hoping to take out the human channel node cell. Ripples caused by the inventions of computer technology cause adverse effects in modern civilization, spawning a novel technology phenomenon.

The past decade motivated hardware and software production tools to improve the network environment. High-speed digital data transmission also enhances the user's ability to process five basic protocols: numbers, text, images, sound, and video. The limited speed and storage capacity of legacy computers, built to handle simple text and numbers, became obsolete. Recently, sound and video telecommunication and computers have emerged as the fastest streaming data transfer demands.

Computer hardware and software has advanced tremendously since the first electronic digital program written in 1946. As of 1994, international standards govern a universal W3C. Early network administrators wrote their own machine instructions tailored to meet organization requirements.

Nature evolves culture evils that are not clear or precise. Technology growth is faster and there is a struggle just to keep it within perspective. However, our task is to manage the evolution and to understand the future influences of the technologies we create. We typically emphasize the speed and influence of technology evolution but overlook the overwhelming role of capitalism and profit. Additionally, unprotected, unauthorized accesses of remote access points provide easy access. The computer industry encourages upgrades to maximize profit that drives technological evolution, for instance. As one scholar Illustrated, "Berners-Lee first introduced the Internet in 1989 at the European Particle Physics Laboratory in Geneva, although no one noticed. A year later, the first browsers invited interested participants." (Gaskin, 2000)

There are those who view VPN knowledge as controlled decision-making force. Several experts agree technology is the dominating factor within an organization and emphasizes its influence. For instance, expertise demands characteristic ways of thinking. Technology enhancements set its own objectives and would have us evaluate progress toward those objectives in terms of its own criteria and logic. These demands and criteria are quite independent of the 'content' of the technology. Technology is more than an expression of culture; technology drives culture. In a real sense, technology is culture." (Wilson, 1989)

The process was swift, personal, intuitive, and collegial; exploring the web has lost age of innocence. Today, billions of dollars, the fate of thousands of companies, and the future of modern civilization rely on Berners-Lee's invention. Furthermore, new national security legislation restricts people's right to privacy of necessity but has overwhelming support, altering its conception of civil liberty in exchange for improved national security intelligence, a key success factor for the global anti-terror campaign.

The rapid growth rate stressed the centralized host table scheme. As is the case for most system problems, design solutions are paramount considerations implemented into a scheme of information confidentiality, network authorization, user authentication, and security scalability. For these important reasons, the evolution of network intelligence is now based upon two technologies-integrated circuits and digital communications. Integrated circuit chips contain transistors that store and process programmed instruction information. Digital communication is today's rapid information distribution process.

A basic understanding of this problem addressed the described issue with import historical benchmarks. The first protocol proved to have shortcomings for linking with other networks, offered inadequate access control and unsecured remote access points, although eventually led to the development of DHCP that does not require an administrator to add an entry for each computer to the database that a server uses. In addition, introduced the improvise mechanism that allows a computer to join new networks by obtaining an IP address without manual anticrime prevention.

Object Management: A recognized problem was information consistency because the file changed too quickly for routine updates. A second area of concern was consideration of hierarchical naming. Perhaps there is no better example for object-oriented middleware than the Common Object Request Broker Architecture (COBRA), often referred to as the Object Management Group (OMG), inspired by interface definitions in the UNIX environment, which permits an entire object to be placed on a server and extends method invocation using the same general approach as described. The design and development trade-offs preserve and enforce object definition rules for the system registers, agents, and server log-on methods. Call worked volumes form a variety of operator service requests to provide forecasts. Work, volume, or call command agents collect historical switch trends.

COBRA focuses on objects instead of procedure from conventional Remote Procedure Protocol (RPC) technology; dynamics achieve widespread acceptance which includes a data representation standard known as eXternal Data Representation (XDR), which rely on Transmission Control Protocol and Internet Protocol (TCP/IP) as well. A set of procedure details (i.e., numbers and types of arguments) use the Interface Definition Language (IDL) to help specify a set of procedures to supply the interface characteristics generating necessary server stubs in compiles of local procedures. In addition, the Open Software Foundation (OSF), which defined the Distributed Computing Environment (DCE), is composed of multiple components and compatible tools, which includes its own remote procedure (DCE/RPC) and permits a correlated accessible procedure, converting data representations. Standard external representation translates local object oriented middleware representations.

Databases or navigational accessed entities are one-to-one or one-to-many relationships, linked to predefined entities at the top of the hierarchy root, which permits limited access to system if available at all. The proposed object database model represents standard SQL-92 relationships between object types.

Collection interface objects in a database of class elements allow an object to contain multiple values of a single property. The collection object identified under the SQL-92 standard includes a set or a list of ordered group of objects of the same type, and an array accessed, inserted, or removed. A Database Management System (DBMS) that adheres to this database standard will support structured objects, function of date, interval, time, and time stamp. (Harrington, 2000)

Proxy Aspect: From a technical perspective, each piece of added software is known as a communication stub or proxy in the client and server. The stubs handle all communications for existing programs. One significant difference arises from invocation because proxies are instantiated at runtime like other objects. A program receives a reference to a remote object, from a local proxy that corresponds to the object. When the program invokes a method on the object, control passes to the local proxy. The proxy then sends a message across the network to the server, which invokes the specified method that returns the results. COBRA makes method invocation for remote and local objects appear identical. In addition, to focus on objects, instead of procedures, differs from conventional technologies. The programmer uses a tool to create procedures when constructing the software program. The software creates a proxy at runtime when necessary. Open network computing remote procedure is often referred to as Sun RPC. (Comer, 2001)

An import part to designing a network infrastructure is the connection of the LAN to external networks such as the Internet. Attaching the internal LAN to the web opens the gate to anyone with Internet access. A potential security risk develops for individual user access, although protecting information through IP addresses remains an important security issue, and Internet connections will leave the clients on the network vulnerable. The internal network is vulnerable at each user connection. Servers send information over the Internet and are vulnerable to improper use and corruption. The uniqueness of computer science method is distinguished by distributed, enhanced, and channeled Internet technology, sensing charge levels for a processor or machine device connected to a computer.

A proxy acts as a gateway for the proxy server and is the single connection to the Internet, as it protects the internal network. Designing the proxy server solution for any network can involve several proxy servers acting in various roles. Furthermore, the basic role remains the protection of the internal network from the Internet through computer ports formed with the IP address created sockets. A firewall protects by closing ports and not allowing requests entering from the Internet. Proxies and firewalls provide similar capabilities although a proxy server acts as the gateway for the outgoing request and a firewall is protecting the internal network by blocking incoming packets. (Simmons, Buse, Halpin, 2000)

Database Aspect: Although database control advancement continues, analysts expect an increase in automation in society to increase unemployment significantly, hinder employee skill training, increase U.S. postage, and raise the cost of virtual living. Fortunately, the demand for electric machine power is reducing the complex energy demands of ambitious electronic devices, such as computers, servers, satellites, wave recorders, broadband radios, and high-definition television. To satisfy the technology needs, a common DBMS offer interesting software solutions of high-level programming convenience without any processing details and a multipurpose SQL search engine.

Today, database application programs specialize in a particular virtual market, enhance revenue resources, or assist in the development of contemporary products and services. Consequently, the number of computer database applications rise rapidly. Nowadays thousands of workstations share common information. Controlled by a central server computer, broadly defined database terms contain many specialized disciplines underlying the production process that enrich the assembly of routine process and share fundamental business goals.

The complexity and cost of implementing managed infrastructure systems has so far hindered major breakthroughs, and cost control of the software development sector is improving only gradually. With a multipurpose SQL engine, merging database technology is reinventing the workforce that compiles the infrastructure data by protecting society from biohazards like anthrax, smallpox, or similar biochemical exposures.

Application programs that access integrity problems and the inability of such processing systems to represent logical relationships of the SNA hierarchical data protocol model provide DBMS support. All databases or navigational channel accessing are predefined relationships, typically appearing in the entity at the top of the hierarchy or route, and proceeds in a high level direct access to data that are limited if available at all. (Appendix, Figure 3)

Server database programs use automated tools to aid in the construction of distributed software. Middleware aspect provides a dependable IDL tool to follow the paradigm that allows the user to specify arguments for each remote procedure. The automated tool then reads IDL descriptions and generates the necessary proxy software automatically. The advantage of the method is the ease of programming techniques, which follow a pattern used to generate client and server procedure transmitted from a computer to a conventional server, then change focus from procedural to object oriented forms.

The large repository of data can be used by departments and users integrated with a minimum amount of duplication. The company's operational data, which stores a description of the data as a "self-describing collection of integrated records" as data-about-data or "meta-data" exchanged from the system catalog dictionary. The self-describing nature of a database provides improved data independence by adding new structures into the continuously modified database of existing structures. (Connolly, Begg, 2000)

Internet Perception: The basic requirements for linking with other networks led to the development of TCP/IP. Even when expressed in 4- or 8-bit decimal numbers delimited by periods, it is a scheme for addressing computers by identifiable names. The Advanced Research Projects Agency Network (ARPANET) and the Department of Defense network are the ancestors of today's Internet, built in the late sixties using a proprietary protocol suite. ARPANET exploded the Internet evolution into the National Information Infrastructure (NII). In addition, the latter uses 32-bit address numbers. Harsh realities of the dynamic environment increased the need for enhanced security, which introduced increased application network vulnerabilities.

Business machines stimulated novel LAN software choices, advancing high-speed digital computers. Such a demand for silicon technology prompted a workforce retraining consideration by skilled business owners who design and build automated WAN systems. In addition, the computerization paradigm enabled development to protect valuable resource information comprised of physically connected energy state technology. Rapid communication technology allows manufacturers and producers to lower delivery cost, analyze how it operates, and increase sales beyond traditional business boundaries.

Given the dynamics of real-world applications, a similar perspective further explained: "Now that the public is given a taste of high-speed broadband Internet access, there's no going back. People aren't just clamoring for speed; they want quality service as well." (Rolfe, 2000) The challenge is to come up with the technologically and economically feasible business solution the system protocols will use. Integrated Services Digital Network (ISDN) developers realized the industry requirements and recommended a Link Access Protocol-D channel (LAP-D), characteristics which are useful tools for several business applications, justification for increasing data transfer speed, and enhancement of digital offerings to provide improved network channel solutions. Technology maturity in communications enables the production mechanism of delivered services availability, how to market them, and the future cast of customer support. Changes in Internet technology knowledge represent future user-defined object operations of automated information distribution control.

Transmission methods are based on the efforts of the Consultative Committee for International Telephone and Telegraph (CCITT) founded in 1988, which is now a recognized International Telecommunications Union (ITU). The perception of the American National Standards Institute (ANSI) provided significant frame relay standards (1990), forming an important foundation which recognized the potential of this evolving technology. Cisco Systems, Digital Equipment Corporation, Stratacom, and Northern Telecom are members of the Frame Relay Forum (FRF) consortium. The focused development trends of frame relay technology facilitate the interoperability of equipment by developing appropriate standards. ANSI and the FRF led to development of the Implementation Agreements (IA), which define frame relay technology transfer rate, frame, data delivery ratio, and service availability for multiple service provider agreements. (Held, 2000)

Information Exchange Heuristics

The advantage of living in the modern age is that computers are much faster, and we need them to assist with memory and disk storage. Corporate America's excitement over technology is cultivating the human productive spirit, although, as network technology evolved, faithful skilled workers have vanished. The educational criteria for today's entry-level jobs has altered the employer's definition of a skilled or unskilled worker and made employee worth more difficult to measure. In addition, corporation's transfer associated risks of building such a massive storage capacity. Client contracts are exchanged, reviewed, and visualized. To gain a better description of the overall process, WAN foundation methods are analyzed for heuristic knowledge then implemented.

A more powerful multiuser operating system was required to run mainframe and PC-type operating systems that interface with a network server workstation, share peripheral devices, communicate with a computer, or a device that stores database information, and assists in the operation of WAN. In addition, a network operating system layer stack is now placed above a primary operating system, such as DOS. The common International Organization for Standardization (OSI) model builds on the seven-layer network architecture (Appendix, Figure 4). Dominant in this category are Novell's UNIX (i.e., an operating system designed for network workstations) and Novell's Netware, the leading operating system of automation machine programming.

The usual response to malfunction in a UNIX operating environment is to call for professional assistance. Unfortunately, unless you are willing to wait on a lucrative preferred client list, ad hoc problem solutions are not processing exclusions. UNIX is a deep operating system, in the sense that it has a history of about thirty years, although security was not a primary design consideration. Intelligent WAN monitoring distribution is a machine control-driven performance of processor speed that accurately tests security integrity. Artificially intelligent business machines enhance support activities, expanding embedded knowledge into four primary categories: personal, mainframe, dedicated, and embedded.

Linking Networks: Automated technology, in addition to the Internet, revolutionized modern civilization by rapidly improving human communication channels or adding novelty to routine business activities. Although the selected data for historical analysis perspectives tend to project a decrease in human contact, it causes a devalued individual spirit. Furthermore, experience has proven reliance on computerized technology cultivates disenfranchised workers. Interestingly enough, the research determined that the Internet has flourished by producing a body of domain name documents and publications following a linked collection of Radio Frequency (RF) connected database sources.

Many argent hackers search the Internet for potential vulnerabilities.

Off the shelf application softwares are not adequately designed to meet specific requirements of the organization. The challenge is to develop the economic player for technology and formulate security methods that offer a reasonable infrastructure protection. The multifaceted IP implemented facilitate routine communication such as "electronic mail" in a personalized collection of routed addresses, full of information packets creating security issues, which hinder the network intruder vulnerability. The exchange in protection of application software development confront security measures and offer administrators a trade-off in information security, data integrity, and confidentiality combined with rapid access dynamics of the environment. The software architecture protocols and utilities allow independent review of system records, test for system errors, and monitor user actions.

Another important concept in distributed linkage technology is common RF communication waves broadcast on radio, television, and wireless satellite channels. LAN host machines are connected through broadband cable television and various Internet services. The network protocols allow switching systems to organize the flow control of transmitted information and authenticate communications.

Work-related experience demonstrates a disruption the system can cause adverse affects and hinder economic survival. Since the WAN implementation venture, the financial risks have increased loss potential and corporate insurance responsibilities have skyrocketed. Because of these important associated facts, electronic information distribution costs offer a profound impact to the foundation of an organization and possibly motivate the recent economic ripple effect on today's business society. Although an intelligent network security initiative is nondiscriminating for system administrators, corrupt information maliciously downloaded from the nternet can harm an organization. The popularity of the VPN is acceptable for the associates and respected clients who rely on trusting technology to preserve and protect the data and resources available.

Rapid Access: Why consider high-speed networking? Are users experiencing bottlenecks in network performance? Possibly, the organization's plan involves bandwidth hungry applications such as videoconference transmission and multimedia applications. Will improved bandwidth actually solve the situation? The answers to these questions require sufficient analysis. The fundamental principle of dependability, adaptability, and authenticity should behave as the intelligent engine of authentication or access control authorization addressing the concerns and requirements of management to protect resources within the system.

High-speed network availability will certainly speed transfers although with escalating costs. The use of high-speed computer technology helps society's skilled professionals break ground to solve today's difficult problems. Visual software tools enable professionals with a graphical perspective on the utilization of DSL advanced hardware and software application.

To launch the proposed network implementation project, a system designer determines appropriate segments that can benefit from high-speed protocol and carefully estimates associated expenses. The analysis should be adequate to determine bandwidth in the near future. Routine upgrades usually extend the system value beyond current financial obligations although are obsolete in approximately two years. For example, several workstations on one cable-extension segment can generate more traffic than available bandwidth can handle. In addition, some applications require even more bandwidth. Although current average packet size is around 512 bytes, the spiraling number may soon grow to 1500 bytes or more. The top 20 percent of bandwidth-starved segments should resolve the problem areas. Implementing with high quality cables at all connections is low in cost and a versatile alternative. Rapid access can improve management sufficiently but there are several alternatives to consider preventing a bandwidth crunch on the segments. The popular bandwidth management tools are virtual networking, microsegmentation, full duplex protocols, servers, hubs, routers, switches, and workstations. (Parnell, 1999)

Rapid access service is vulnerable to economic rivals, organized crime, and general purpose pranks, causing alarming impacts on resources. The financial risks of some portions of the intelligent information infrastructure are enormous. Furthermore, an incorrect router setting can allow IP leakage that leads to unauthorized access and provides the intruder valuable DNS information. Unsecured and unmonitored remote access points provide easy access to the VPN. Without a correctly configured firewall, the unauthenticated intruder service access is granted. A database becomes vulnerable in the sensitive relationship environment that allows greater threat to security. The alternative is intensive testing to detect system errors and automatically collect corrupt presence, then correct the problem.

Energizing the WAN environment requires more bandwidth. Until every office and home has bandwidth, passing knowledge across the Internet requires careful planning about the goals of the user training and focused attentions of the organization. Academia-infused training may follow a traditional strategy although corporations usually have more money, bandwidth, and autonomy. By 2004, the FTP standards motivated organizations to adopt common tracked recourses, eliminate redundant management objectives, and remove system-learning obstacles.

The rapid-access enterprise has come to be the popular description of the new principle that time and space is no longer the organizing foundations that they once were. A virtual organization does not exist in one place or perhaps, even one time. It coexists whenever and wherever the purchase of goods or services happens to be. Today's infrastructure organizations may actually have communities of common practice. (Sprague, 1998)

Resource Channels: The use of tools, power, energy, and materials are generally feasible improvements for the purposes of production. Almost every human process for information protection of applications software and security depends on complex technological systems. Technology covers a broad spectrum with associated science applications. The progress suggests it is a "continuum stretching from very basic scientific research, through applied research and technology having diverse social, economic, and political implications;" many scientists hold this theory alike. (DeVore, 1992)

The negative effects of channel technology do not outweigh the positive potential benefits of an automated society. Visibly, an improved social benefit is increasing human technological opportunity. Now, we see the future as a highly technical world focused on the high school graduate with assumed skills in computer technology. Policy intervention means a demand for knowledgeable network technology professionals. Government educational requirements for professionals alter security strategy procedures, modeling the potential of twenty-first century intelligent Internet business information systems.

The design techniques to cost-effective network availability are the basic approaches of high availability network design. Defining essential commands offer a full range of network management capabilities and networking protocols or offer more bridging and routing alternatives. The International Organization for Standardization's Open System Interchange (ISO OSI) reference model definitions of connectivity offers various communication protocols to enhance the WAN design. Commonly, the TCP/IP networks interact with the OSI seven layer international standards that are a defined stack reference model of normal system architecture. The OSI reference model also defines four classes of intermediary devices: repeater, bridge, intermediate system, and protocol conversion gateway. The purpose of OSI in business applications is clear compatibility options, which provide a common physical link between network systems. OSI standards actually communicate and negotiate through intermediaries and devices referenced within the seven layer stack reference model. (Jones, 2000)

Resource channel management is often offset by a multifaceted discipline encompassing a wide range of management and technological specialties. Information specialists agree resource knowledge will help achieve the necessary goals. Furthermore, they accommodate the needs of a VPN society, forecasting sensible organization to an advancement of WAN discipline. Resource channels prospered tremendously through parties sharing a similar concern and cooperating.

IP Address Routing: The IP assignment lease can be permanent or valid for an interval lease time. Devices with nonvolatile storage attempt to renew their leases with the DHCP server upon start-up. DHCP represented a tremendous time-saving advancement for network administrators, who previously visited each device to configure its IP address and other parameters, the default router, and named servers. DHCP devices obtain information out of the box, without any prior configuration; DHCP need only be enabled. From an IP address management perspective, however, DHCP servers represented a new class of labeled servers to configure and manage.

Configuration is similar to DHCP lease pool subnets automatically and efficiently controlling IP addresses, such as the name-space service directory, static addresses in spreadsheets, or bundled server packages. Address conflicts hinder the network, monopolizing bandwidth of the dynamic IP management. Dynamic IP addressing integrates DHCP, DNS, and other IP services through a single interface. Administered, fail-safe addressing eliminates the potential for conflicts. Fault-tolerant services prevent redundant capability.

The common DNS server implementation is mandated by the rapid growth and dynamic nature of today's private Intranets and the public Internet. DNS solutions in today's networks increasingly rely on DHCP for dynamic IP address assignment. Windows 2000 XP and Active Directory (1999) are server software implementations that address the future of dynamic, policy-based networking interoperability with traditional DNS implementations.

For example, the symbol 10 in the binary system corresponds to two in the decimal system. Combinations of bit charges represent numbers, letters, portions of pictures, and sounds—all the data that the computer processes and instructions to process the data. Millions of tiny electronic circuits like an ordinary light switch are measured in multiple bytes.

Potential data enters the system as electrical charges that represent numbers two levels of charge. Computer networks deliver bits of information from one point to another. One requirement for transmitting information is that the computer systems on each end speak the same language or protocol. For instance, to address a regular letter or package in the United States, you need to write the destination address on the front of the envelope. Similarly, IP-to-IP addresses specify both the source and destination systems on a TCP/IP network. Each address consists of 32 bits, usually broken into four decimal numbers separated by dots. Each decimal number represents an 8-bit byte (an octet) in the address. The case in point: 00001001 01000011 00100110 00000001 32-bit address

9. 67. 38.1, a decimal address, so each address is separated into logical parts. The network address identifies the region (or section) of the total network that contains the subsystem.

The machine host address is similar to an apartment or suite number because it specifically identifies a particular system within that region. IP addresses belong to four classes, depending on the 32-bit class address: split, fifth, A, or E, which is not common. Class A addresses use 7 bits for the network address portion and a 24-bit host address for 126 (27-2) possible networks (regions) and over two billion addresses. One bit identifies the address as Class A, to distinguish it from other classes. The IP subnet mask assigns a Class A address for each WAN layer.

Technology Standards: Guidelines developed by the National Institute of Standards in Technology (NIST) plays an important role in setting computing and communications standards. NIST's Advanced Technology Program offers financial assistance to companies developing technologically advanced products during the research and development period, and additionally played a considerable role in nurturing the US computer industry guidelines.

A sensible America accommodates the influx of technically exchanged trust relationships which helps resolve information leakage issues or intruder operating system attacks vulnerable to application versions, shares, groups, DNS information zones, or running host services such as, SMTP, FTP, HTTP, HTTPS, POP3, Restrict Telnet, Sun RPC, NetBIOS, and Pass-Out, Translate-All remote access points, therefore improving the information transfer process.

Additional standards are designed to carry voice traffic digital signals that facilitate data traffic through the telephone line based on federally regulated long-distance connections. Leased circuits provide a digital circuit that extends between two buildings, across a large city or from one city to a building in another. Connection cost is based on the capacity of the circuit and distance. Digital common carriers form the fundamental building blocks for long-distance networks. Leased digital circuits must agree to follow rules of the FCC including standards that were designed for transmission of digital information. The computer industry and the telephone industry are independent entities with individual standards.

A special piece of hardware is required to interface a computer to a digital circuit connected through telephone service. The Data Service Unit (DSU) and Channel Service Unit (DSU/CSU) contains two functional parts, usually combined into a single chassis required at each of the DSU/CSU leased digital circuit that converts between the digital standards used in the telephones and computer channel vendors. (Appendix, Figure 5)

Radio Frequency Properties: A network that uses electromagnetic RF transmissions does not require direct physical connections between computers. RF technology can be combined with satellites to provide Geostationary Earth Orbit (GEO) communication across longer distances. The satellite contains a transponder that consists of a radio receiver and transmitter, which amplifies RF and transmits the amplified signal back toward the ground at a slightly different angle. Multiple transponders operate independently, typically six to twelve, and use an individual RF (i.e., channel). Multiple Asymmetric Digital Subscriber Line (ADSL) resource transmissions provide service simultaneously, can be shared or serve many clients.

A Discrete Multi Tone (DMT) modulation combines frequency and inverse multiplexing technology. To accommodate differences in local loop characteristics, ADSL is adaptive and probes the line between them to follow a recognized characteristic and negotiate optimal communication for the line by dividing the bandwidth into 286 separate frequencies or sub channels, with 255 frequencies used for downstream data transmissions and thirty-one for upstream date transmissions. The common two channels are reserved for control information stored on the SUN Enterprise 250 server in the standard COBRA infrastructure.

To maintain a reliable connection a separate modem runs on each sub channel and the modulated carrier space is configured at 4.1325 KHz to prevent signal interference and avoid using the bandwidth below 4 kHz, a modulation scheme that includes fewer bits per baud. Conventional twisted pair wiring permits a robust technology that adapts to various conditions automatically, does not guarantee a data transfer rate, and performs coexisting specifications. The downstream rate varies from 32 Kbps to 64 Mbps and permits two modems to access the select optimal frequencies as well as modulation techniques. Other similar and important DSL technologies are available (e.g., SDSL, HDSL, and VDSL).

Engineers have frequently devised ways to use the existing Cable TV (CATV) infrastructure as a local technology that delivers digital data to subscribers. Large bandwidth in systems is not sufficient to handle frequency division multiplexing schemes that extend to each user. The original cable infrastructure theory provides downstream delivery and could not provide two-way digital communications, which is a serious disadvantage. Engineers continue to explore several methods to resolve the issue of downstream time division multiplexing. A new paradigm for delivering high availability within the VPN industry stands poised upon the threshold of a new era in which smart devices as switches and routers will become integral parts of the Internet's infrastructure. (Sheaffer, 2000)

Significance of Automation

The true definition of intelligent automation is not clear. The important questions are: What is the purpose of expensive automation? Can enhanced technology reduce production costs and improve client service? Mechanization is reducing risk exposure to worker liability losses, and technocrats insist that computerization increases quality leisure time to help civilization perform routine activities; but automation can force a working society into early retirement or result in a novel career choice. The traditional human boundaries are crossed, and the employees who remain are those willing to learn new computer technology skills.

So what can the system designer or software developer do to satisfy both internal and external segments of management? System development presents complex trade-off issues that provide anticipated balancing for feasible fix protections. The organization must thrive to survive and anticipate corporate growth. The key question to resolve is clearly, what can VPN potential provide the company?

One professor's perspective insists technology does not cause unemployment; it increases the amount of work that can be done and claims the rulers of the institution must devise meaningful activities for the people they lead. Expert system designers agree training and development is a major component of system implementation. The level and duration of training depend on the user's knowledge level of the system attributes. (Awad, 1996)

Another perspective rejects a scientific solution of business problems and foresees the collapse of technology. The literature of doom is best described in *Decline of the West* (Spengler, 1922). Management purists demand human resource solutions although reduced labor organizations are a positive attribute of machine-driven business operations. The office goof-off no longer exists, as activities are monitored and recorded in an administrative statistical analysis report.

One consensus viewpoint insists that the consequence of scientific technology improvement is beneficial, although scientists often refuse to share their perspectives without financial rewards and are seldom eager to prescribe a permanent cure. Solutions for the deployment of automation presume science's dominance, but objective science does not make exaggerated claims about technology.

Automation is best described as "a set of various words in the set of ideas that was a shortened form of an earlier phrase, automatic operation." (Garth p. 85) Scientific progress is devoted to systematic possibilities and often fails to measure the social consequences. The assumption is anything that can be done must be done. Mythology, a strong social and political force, guides the problematic future of the technocratic expertise, although information infrastructure security depends on technology which plagues the industry and is mandatory.

Another scholarly consensus argues automated machines improve work capacity, so what is the purpose of doing more work? Pseudo technology advancements are not a reliable practice and are a waste of natural resources; they convert short-lived merchandise into a wasteland of mass production and consumption. A lack of acceptable resource policies prohibits enhancement opportunity to foresee the widespread cyber-waste advancement discovered through a genuine desolation.

The pervasive deployment of these devices carries with it a requirement for High Availability (HA) performance to meet market expectations for continuous Internet access or uptime. Measuring significance of the battle as well as determining adequate tools plagues the development. Unfortunately, society is frequently left to recover from pseudoscience fads in a frequently altering industry.

Educational Issues

A realistic consequence of worthy facts reveals automated technology has forced talented people out of work and technologically skilled reemployment is perhaps the only cure. Employment requirements frequently change and departments revise their thinking, as workers are isolated or eventually relocated. Unemployment caused by a lack of technical skill hinders society and frequently delays reemployment. Requiring computer skills before employment creates obstacles and determines whether the job is even available.

Employees are shifted from old to new forms of employment, with the least possible pain to both themselves and society. Information can be processed with astonishing speed and accuracy with skilled workers.

Furthermore, a primary training barrier is expanding the range of production output to a virtual class from management's perspective, facing considerable resistance to changed opposition. A potential class of talented workers questions administrators intelligently, offers managerial expertise and executive decision making, and is highly capable of channeling a central control system of integrated conventional processes. The government of today's VPN industry is both latent and organized in terms of ideological obstacles and a formidable political opposition.

A recent automation plant closure resulted in several months of problems associated with unemployed workers. The plan focused on pre-shutdown planning and continuing education, job placement and relocation, and negotiated contract incentives. Some workers were eligible for pensions and successful training objectives. Financial exposures resulted in severance pay, job training, and personal placement consultation. The average amount of time for job replacement was six to nine months.

Labor unions argue the legitimacy of the cross-training problem, although modern computerization has proven to be more efficient, less time consuming, channeled administrative control, and provided enhanced standards of survival. Responsible workers in society have a legitimate claim to sources of support and a viable way of existence. In times of economic crisis, numerous advantages of computer applications accurately assist management decisions, offer effective task applications, and control the net worth of business profits with less economic bureaucracy.

Additional enhancements in technology training offer no guarantee of increased profits or anything tangible to produce substantial income. The actual market value is based upon supply scarcity and public mandate. A realistic reservation is reliable resources to fund and manage the growing historical mania. Actually, there is not enough financial fuel for the recent technological pursuit.

The system training is assured when values and beliefs reflect the popular civil technology. The pragmatic movement to solve technological problems moves forward with enormous emphasis on leadership, well aware and optimistic, concerning a probability of success. A confidence that scientific progress enhances well-being and enrichment to fulfill the guidance objectives cannot enlist the talented alone. Additionally, there are varieties of factors that can affect the success of a technology-based training program. In general, they can be divided into three broad components: organizational factors, environmental factors, and individual factors. (Noe, Ford, 1992)

Perfect systems alternatives usually are not feasible resource trade-offs so designs offer a variety of viable systems choices that adequately addresses certain object types of definition control. Processes are sequential and largely discreet steps built upon substantial research in scheme development oriented toward control of new inputs, outputs, defined security, and catalog moments. Historically processes are defined in terms of work for tasks and activities transforming inputs into outputs. Business process inside are to remove cost through improved system performance. The input process output is a powerful conception although limited. Total Quality Management (TQM) reengineering strongly suggests a viable design discipline. (Keen, McDonald, 2000)

Total Quality Management: An alternative is to change insights and practical network solutions, which will focus on the importance of both invention and inspire commerce focus. TQM strongly suggests engineering discipline that began with Frederick Taylor's work in the 1920s and inspired the quality movement reengineering revolution or possibly the disruption society depending on your perspective.

The implementation of automated technology requires greater technical skill, which can thus demand higher employee wages; it also requires a genuine desire to work behind a computer. The chief programmer mandates reliable training of administrative technician's complex specifications, which can be misunderstood within the scope of the model suitable for general work.

Machine topologies are designed according to what the machines will do and how they work. Machine balances are cultivated varieties of the system tools, which amplify ability to enhance what ever we can already do. Proxies can replace us by doing what we already do perhaps better, such as a computer program that integrates mathematical functions. The novelties of a system perform tasks that we could never do without their assistance, as in space rocket technology. Employee involvement adds importance to the institution.

A variety of educational disciplines are a common outgrowth of implementing automation The system managers remain important producing part of a product but correctly programmed machines replace workers who are also replaced by multitasking associates with added responsibility and job-related stress. Human dislocations necessitate retraining although some employees break under the pressure.

Mechanization engineers are pioneers driving self-protection and adding bargaining power, directly affecting the proportion of organized workers. Technicians lose membership and financial support if the machine becomes more valuable than the employee does. Achievement of the information challenge obstacles helps understands the heuristic variables of intangible information and the tangible medium of distribution. Cohesion in terms of the appreciation is the tangible accomplishment of Western society.

"There are signs that with the advancement of automation, we are beginning to lower production time and dealing with leisure as a civilized human should." Recent indicators describe the advantage of automation as allowing for "civilized leisure." Indulging in arts and intellectual pursuits is a learned behavior that permits humanity to fully enjoy life and push itself to a recognized limit, bringing us closer to human relationships. (Pyke, 1957)

Environment Viewpoint: Realistically, employees being replaced by a skeleton crew are a threat and a social concern. As a result, fewer industry associates are motivated to reshape the business environment. Fluctuations in the new scale of demand address the need to compensate hardships of readjustment. The new breed of management requires more education but entertains a sense of satisfaction for creativity.

The phenomena of technocratic advancement may soon cause a post-modern disaster. Automation solution enhancements cannot exist alone amidst a machine-driven humanity. In the early days of computers, punch cards were fed to keypunch operators who checked the specifics of card thickness and sorted them into a special machine file folder of rectangular holes in order to perform a specific task. An electronic typewriter improved keying and eventually inspired printing and editing of a document, which proved faster and more efficient for repeated postal activities. By the late 1970s, word processors had become accepted, although they were unable to host electronic improvements. Law offices and accounting firms benefited with computerized case preparation and secretarial work was significantly reduced. Standard forms required information entered by the user to create information database containers of registration assembly code that seldom changed.

Computer science significantly reduced the size of expensive mainframes, offering increased storage, reduced price, and feature-rich software. Applications included word processing, spreadsheets, and a database of client's personal information. Eventually, the fixed drive improved storage capacity, allowing creation of larger programs.

The word processing programs eliminated redundant typing but were slow and cumbersome for the average user. On the other hand, spreadsheet programs offered affordable business solutions to create a budgeting proposal or accurately project the project's expenditure. The acquisition of information allowed administrators to evaluate alternative possibilities. Inventory tracking, product information, and customer profiles offered valuable data mining statistics encrypted, sorted into a menu and accessible at different levels of context.

Finding a climate-controlled room to store sensitive equipment was an unpleasant task. The average employee of the earlier period was not prepared for skilled-mind work. Computers seemed to represent a thinking mind and an independent entity. Frequently, humans blame their problems and mistakes on computer error. Somehow, workers believe the programmer is not responsible for computer errors and the machine is always defective.

The average business machine has dramatically improved over the years, tracking electronic magazines with browsers on the Internet. The early years of computerization demonstrated that the expense of implementing large-scale computer operations was too great for the average business to control. The price tag associated with network technology operations could not justify the financial prosperity it hinted at.

Eventually, a database centralized location allowed multiple user access to the same information. Early program engines could not perform difficult computer task activities with intelligence because of substantial speed reduction insufficient for rapid digital processing.

Numerous users share the anxiety about computers because they have had bad experiences, were pushed the wrong way or possibly deleted files accidentally. Computer solutions usually are teaching the accessing of specific task commands and help input data without complicated functions or unfamiliar software. This concept of education has not changed dramatically.

Nowadays, the average word processing program is user-friendly and reasonably easy to operate. Functions are programmed through recording functions and macro keys, reducing repetitive data input to several tasks. A macro helps the user run a series of programmed steps with a few keystrokes. (Appendix, Table 3)

Software is regularly updated to increase speed and accuracy and decrease user deficiency. A database program is modified to add columns of defined categories necessary to track tax consequences and penalties as well as to present reference tables for short- or long-term assessment. A flexible database application tracks customers by name, address, contact information, purchasing patterns, and personnel qualifications, allowing visualization of business peak periods. Reordering, purchasing, and security alerts can be programmed into the system.

An integrated package offering smooth marketing preparation management tools helps recognize patterns on demand. The decision to computerize operations depends largely on the complexity of environment and pertinent government regulations. Automating a business means identifying what the computer can do for that business. A basic blueprint of customer alternatives helps individualize hardware and software. Tracking account inventory and financial statistics are compiled for future marketing strategies.

The increasing choice of useful software, hardware, and application causes confusion for an organization trying to determine what it needs. The electronic process edge allows organization to extend the opportunity to create markets, both within and beyond the scope of traditional boundaries.

We are entering the knowledge age in which the basic economic resource is no longer capital, or labor, but knowledge. Human resources play a central role in a fundamental shaping for future business opportunities, although the largest companies in the country have downsized in part because of a broader restructuring of the industry. A paradigm shift affects not only business but also society as a whole.

Organizations focus on their core activities, although outsource work to specialized companies. This trend is associated with a version of business activity and growth of self-containment. Rapidly changing patterns of business communications will have a profound effect on occupation, knowledge based products and services. Knowledge is intangible although presents a high value to organization.

Unlike people, computers are emotionless collections of cyber nerves that can stay connected twenty-four hours a day without intervention. Workers require compensation, sick days, unions, and improved health benefits. Successful resource consultation inspires profit sharing and on-the-job training. The new economy shift is to service-related media and is telecommunications sensitive. Collaborative technologies are a tool for augmenting human cognition. Successful strategies will exploit the developments in technology and take full advantage of the e-commerce to communicate new products and services. (Skyrme, 1999)

Prevailing Knowledge Significance

Computer trends are changing the shape, size, and boundaries of communication devices by adding mobility and ease of use. Today's wireless phones and handheld computers are equipped with operating systems similar to PCs. The increased use of the Internet and WAN-connected computers stimulates recent trends in telecommunicating and increase the significance of the personal computer growth magnitude for the community. The speed and power of computers and the Internet have also improved dramatically over the past few years. Consequently, software has become easier to operate and network users enjoy more leisure time and a higher rate of productivity.

A boundary between machines and humans, and science and nonscience, has varied from time to time and changed our environment. The challenge is exploring and demonstrating the permeability of these boundaries that reveal our understanding of how the environment is a constituent factor relative to the way actions shape the economy. From one perspective, humans and machines act in a machinelike fashion. Does the possibility of emphasis in the boundaries between humans and machines exist? All our employees are willing to change reasoning for acting similar to a machinelike fashion. Exploring the possibility shares the position of a boundary between human and machines or the change in which they act. Can the motivating factors of machine management alter the way employees communicate between themselves and the intelligent machine? The requirements of labor personnel are not fully realized, as skills and applications in technology are advancing at an accelerated rate.

Embedding Components: Mechanized technological environmental instability, or the major dichotomy advancement of shaped intelligent actions, provide a focus of understanding to the processes of mechanized automation to cultivate the consequences of what we have proposed. The study of intelligent machine design is an order of development to a top quality performer based on what machines will do for the organization interwoven on how machine and resource motivation improves work. Determining the differences between the types of machines discussed in the VPN theory is a radical difference and elaboration of the economy models, which estimates actions as examples drawn from administrative network experience. A series of definition actions in the methodology assumes a presumption in the approach to automation output, which is a feasible embedded solution, and speeds up routine operations by reduction of lower skilled personnel. Impending changes are more dramatic control devices to improve production and decrease the need for labor.

Ad hoc decision success or failure dependence is based on feasible truth or proof of resulting consequence. Describing what we did, using and comparing expression is the concept of "closure" and helps develop a number of pattern concepts and a relationship for "black boxing"—the application to the concept of "microworld" perfection. The crucial difference between behavioral actions of human resources and anticipated intelligent machines is the adequate understanding of historical development defining through black box experimentation.

Technology and science are much the same from one perspective so the common sense retraining difference is between science and technology "black boxing" delegation of responsibility to machinery or description of relative actions. Intelligent machines are our "collective" proxy server, and poor linguistic performance continually reminds us of the instruments aptitude of our linguistic abilities. (Collins, Kusch, 1998)

Collaboration: Sharing and exchanged business information is a vital discipline helping to forecast the future course for success in a fluid business environment. E-commerce is energizing business realities and emerging the new economy; furthermore, the fundamentals about customer relations presents a new brand of service dilemma, which is identified as a critical factor of intelligent reasoning through described expert approaches. Steps are necessary to improve efficiency and combine Internet technology knowledge invention into execution, as a larger motivating force, leaving people with more initiative, and measuring the Internet economy extensive overvaluation as a campaign of financial controversy.

A focus of electronic process motivation transforms inputs into outputs. Motivation incentives will not decrease worker responsibility although historically, processes are defined terms of work "tasks" and "activities." Processed sequential timed output results transmit to discreet policy defined steps of resolved objective business definition. Cost containment inside the organization is to remove detriment through improved performance. The systems input-process-output is a powerful conception although limited to a traditional view process as illustrated. (Appendix, Figure 6)

Commerce processing edge is built on business relationships and is coordinated through the interface of exchanged value or a direct contact through TCP/IP. The Web site software is the central flow control of information and processed customer requests, in addition to defined response. The process is responsible for transferring the input "request" into the output "response" link through a set of tasks and activities that constitute a linear flow of sequenced steps. Value judgments can be rules regulated through software's electric linkage of workflows or business rules that govern the transformation of request and response. Thousands of people who are aware of system vulnerability and act to protect firewall security protocol recognize achieving information proficiency.

Business rules may overlook the importance of processes in e-commerce combining features and processes described about workflows, steps, tasks, and activities commonly used for ordering, inventory management, shipping, pricing, and payment. Rules are the statement that donates the interaction between users.

The value in e-process WAN technology is a recognized winning of customer and leveraging assets, which allow for customer approval and extends the revenue share perspective. Executives who leverage the value of WAN concentrate their management attention to capital identity capabilities, which use options to change the system structure and move companies towards virtual integration through embedded system rules. Software development advantage of customers and suppliers increases knowledge base by reducing the translation activities.

Tasking application software leverages the best capabilities available in the market and reduces risk exposure or redistributes business risks to suppliers, intermediaries, and income mentors. Outsourcing by using technology and the electronic process principles strengthens topology model operation in the direction of redistributed business rules, which make accurate decisions and improve coordination, offering greater efficiency. Handling exceptions secures a future perspective capable of sensed response to customer needs with ability to perform the unusual situation, often the untapped source of customer relationships and invention. Fear, compounded by uncertainty, have paralyzed companies, created process blindness, and allowed companies to move information noise. The future challenge is to cut through uncertain influences and act with fundamentals that will continue to form the basis for our success. (Keen, McDonald, 2000)

Chapter Summary

The aggregate of benchmark discovery reveals that major advancements in automated technology have occurred in the last one hundred years, and that the previous fifty years of technological invention significantly advanced the intelligent mechanism. Important WAN components are discussed with key fundamental properties, topologies, and terminologies introduced. Computer science protocols, combined with future goals of transfer speed and resource processing distribution, are directly linked to sensing nodes and cables, infrared light signals, or low power RF transmissions over short distances. LANs are connected through TCP/IP lines, television cables, or DSL, T1, and frame cloud satellite links, and offer viable DNS DHCP solution methods to WAN administration. Machine obsolescence and resource training is a significant consideration for system designers, although trends distribute establishment power to the technocratic elite.

A network of networks channel routed layers of automation intelligence stimulates the organizational model of affordable and feasible WAN business solutions, although it increases the threat of intangible system misuse and abuse. The session channel layer of embedded environment components interact between human interventions and a coexisting intelligent exchange progression of COBRA and OSI applications, distributed on a growing UNIX/Windows client-server-host machine collaboration. However, implementation of automation does present negative possibilities. Skilled WAN employees demand higher salary, frequently eluding solutions of diverse human resource training. Resource consultation inspires profit sharing and on-the-job training. Impending changes are more dramatic as knowledge engineers evolve new control devices to improve production possibilities and decrease the need for labor.

Augmented automated business machines mean increased pressures for cross training and unemployment claims. Disfranchised human resource motivational requirements are now visualized standards of machine retraining objectives. Ambiguities of employee obligation, along with recent broadened national security intelligence, represent novel privacy and risk exposure liabilities and motivate expanding system security provisions.

The most efficient VPN transmits not only voice but also digital processes of security measures identifying system misuse and management priorities, and reveals important trade-offs in design and resource development. Computer enhancement builds on the novel process of mechanization to improve manufacturing and production efficiency. In addition, sophisticated programmable machines, rather than improving the productivity of manual labor, have overpowered human resource management. Correct software tools are designed to improve and monitor scheduling, review channel distribution, design architecture process, and evaluate administrative WAN standards.

The review of the literature also reveals effective tracking methods, improved process speed, and motivating e-process technology that can be VPN energized for resource management. WAN operation procedures and intelligent testing software tools decrease outsourced production costs. Clever information systems can accurately forecast the growth of the organization, diagnose system problems, take affective corrective measures, and help administrators determine training requirements. Thoroughly testing standards or comparing solution software alternatives provides managed insight into future financial distributions, production goals, and security discipline precautions. The best defense is the executive who leverages the value of WAN by concentrating their management attention to capital identity capabilities, which use intelligent options to change the system structure and move companies towards virtual integration through embedded heuristic knowledge. E-process leveraging of channeled customers and suppliers increases database activity and system integrity reducing the exchange translation activities.

Chapter 3

Description of Methods

Application Tools

This chapter will describe the methodology used in the VPN design implementation along with the firewall security considerations to provide database analysis. Information input is based on the review of literature findings and the researcher's hardware and software reengineering application experience, which is the basis for the review. The study evaluates routine business operations with design consideration based upon UNIX, NT / 2000 / XP, and CORBA architecture environment. Experiments are conducted based upon shells written and structured for high-level programming of C++ source code and embedded security precautions or utility functions. Mackie real-time automation is applied to enhance routine operations of and examine computerized multichannel development. Application program interfaces are described in detail and provide import transfer and security protocol considerations. Major settings and their applied application programming interfaces are discussed. The amount of detail done through manual labor in the average VPN office has lessened or been eliminated with personal computer workstations.

The training considerations of the computerization project discovered problems and issues that distinguished traditional academia limitations of knowledge authentication. Compiling database into a heuristic decision-making strategy is an import problem-solving tool of reasoning analysis of a human experience, problematic to a specific domain, or the means for determining the validity, uniqueness, or limitations of the study that are summarized. The final implemented VPN system and user definitions are illustrated.

Approach

The researchers UNIX, ANSI C, and Windows reengineering project is chosen for the study. It covers the reviewed approach, database analysis of the study, WAN design considerations, hardware and software components, and descriptions of firewall interface protocols, firewall security strategy, and IP channel address. The ANSI C source code is integrated with Windows 2000 / XP multimedia characteristics based on digital three-tier database architecture and application extends the average computerized network. (Appendix, Figure 7) Consultants agree that a reengineering the project is warranted for a group of workstation computers connected by telephone lines, satellite communications, or RJ45 (i.e., Twisted Pair) cable into a SQL database hierarchy.

The implementation methods and security strategy considerations observed throughout the application discuss implementation issues and main objectives used as a guideline for WAN analysis. Investments in technology are typically pricey but highly advanced and stringently copyright-protected compared to applications of the past.

Reviewed terminology and data object representation determines the application function and necessary system heuristics. This knowledge is not well-defined or techniques simple to distinguish. Algorithm techniques are important computer program theory, although expert systems are designed by explaining the knowledge reasoning of determined results. RF technologies offer unique features, although increase requisite for reliable security solutions. The tragedy in America increased the demand for monitoring the network environment. In addition, recent tragic events in the nation are causing a number of network environment concerns, and government regulation will certainly revolutionize automatic channeling intelligence.

Finally, the VPN review will focus on system applications and software management requirements including electronic commerce marketing needs, wireless vulnerability, and approach to training, customer service administrative policies issues, and macro development of the COBRA Graphical User Interface (GUI).

Description—(1) microprocessors, (2) memory chips, (3) input devices, (4) storage devices, and (5) output devices. RF applications approach, intelligent embedded systems—machine control for network multimedia devices; Frame relay standard—packet switching protocol; 2Mpbs, digital word, IP channel.

Data Gathering Methods

The Microsoft Office Professional software tools can determine meaning for an assortment of business information. Access and Excel are the software tools used in tracking research data examined for review and they suggest statistical results for resource analysis. Integrated invoicing modules distribute financial data model and generate accounts payable and receivable registry tables. A flexible user interface calculates user negotiated rates based on the following criteria: client applicant type, specialized ability, and client-rule based tables. Automatic payroll generation is calculated from multiuser infrastructure exchange. Security system administrators export payroll and billing data to office software packages and tracking schedules are directly linked to staffing and client scheduling.

Customer information is maintained by using the query theory to add and delete in a user-friendly window, and capturing is keyed to a client specialized area, availability, employee classification, restrictions, orientation, and defined context information. User activity is monitored within the scope of the system, measuring staff productivity, and entered directly into scheduling links, capturing details of client and employee information exchange.

Customized data fields are context specific flow control into system administrative analysis and can include scanned documents, contracts, and project benchmark reports. Paper does not exist except on laser or dot matrix printers. Tradition page forms are designed for staffing, billing, and payroll. Data input TCP/IP connection is intelligently automated to a specific customer response. Integration of business through VPN technology is enhancing responsiveness to the technical challenge of the automatic sensation. The efficiency of time-insensitive manual labor is reduced and reports are compiled daily matching experience and credentials with facility insight and productivity.

The database application software is reengineered to a more flexible three-tier COBRA platform to address a customer's diverse concerns branching into new available regional markets, as well as training users to use the new GUI. (Appendix, Figure 8)

The UNIX command shell permits log-on to the system although it opens the door to mistreatment and presents the possibility of deliberate or accidental events. The server database system is distributed to machine controlled point-and-click custom paging. Automatic flags alert the user for duplication redundancy reduction. Automatic tabled rate adjustments are useful for tracking employee overtime calculations. The Cisco routers are scalable technology used to isolate multiuser configurations and adapt to the VPN environment.

VPN Descriptions

Evaluations of business standards result in several administrative directives such as providing a VPN through an intrusion-free gateway into high-speed WAN technology. The experiments were done throughout the researched investigation of WAN application reengineering to determine integrated electronic commerce solutions and automatic information data exchange to users. Computers are indispensable to the routine business environment in modern society. Anticipated procedure helps the organization scan or store bar codes, print with a sensor, and automatically tally the customer bill. Credit card purchases are processed almost instantly. Today's rapid megahertz machines are easier to use and unleash the Internet to the mainstream.

The LAN common thread to automating WAN is expanding a limited network potential. For example, a typical Local Area Network (LAN) consists of two or more personal computers, printers, and high-capacity disk-storage device that enables each computer on the network to access a common set of files. LAN operating software interprets input sensors as electronic triggers for turning communications on or off, sharing printers, storage equipment; and distributing channels simultaneously. A LAN channels multiple architectures into a gateway, converting data as it passes between with systems. WANs connect various computer workstations and smaller networks to larger systems over greater geographical areas. (Appendix, Figure 9)

Routers: Computer channels connect cables, optical fibers, or satellites commonly accessed by modem frame relay synchronized to the Internet to link millions of computer users. UNIX ANSI C software controls operations, directs Cisco routing of ISDN Primary Rate Interface (PRI), and coordinates table-processing rules. The network modules provide ports for connection to a 10- or 100-Mbps Ethernet LAN and to an ISDN PRI LAN. Straight-through, two-pair Category 5 Unshielded Twisted Pair (UTP) color-coded cables connect the RJ-45 port on the Fast Ethernet PRI network module to a switch, TP-hub, repeater, server, or other network device. Computer keyboards or common mouse commands manipulate smooth transition segues. The transmission delay, wait, and tolerance for error are critical factors for delivering products or distributing Internet services. The cable extension method of connection has a maximum throughput limitation. Hubs and repeaters are non-intelligent devices. The port is amplified and sent to all ports. Network slowdown is due to cable segments on the network. One network transmission fills upflow into all cable segments in one active state. There is a variety of positioning mechanism limitations. (Appendix A, Figure 10)

Wireless Mobility: Wireless capabilities present security problems for the Internet, Security, Applications, Authentication, and Cryptography (ISAAC). Digital data is vulnerable to intruder cloning. A recent research group study in Berkeley reported that wireless LANs have multiple vulnerabilities, including a susceptibility to passive attacks aimed at encrypting traffic based on statistical analysis; a process made easier by the broadcast nature of wireless systems. Wireless Encrypting Protocol (WEP) flaws may not detect unauthorized traffic.

Telecommunications intelligence evolves as wireless mobility expands potential clients of W3C. Reliable security measures prevent otherwise inevitable fraud. A sophisticated intrusion detection system for the Internet echoes network protection authentication. Sophisticated sensors monitor network disturbance and detect digital footprints. Hackers and intruders cause mayhem to an organization, hindering channel distribution. Management depends on Internet technology, continually testing data protection resources. System security communicates e-mail resource management protection feedback of transmitted channel data.

Data recognition automatically converts text, data, and encryption compression. The user interface router connects channel layers of computers, verifying user authenticity and establishing a secure, frame cloud gateway of communication. Files are compressed, decompressed, and stored in encrypted database tables and system files. Besides physically connecting workstations, computers, and numerous communication devices, the virtual enterprise is establishing a cohesive architecture, allowing a variety of business machines to transfer simultaneous information in a near seamless fashion.

RF channels connect cables, optical fibers, or satellites commonly accessed by modem and routed into a WAN gateway synchronized to the Internet. A network-to-network operating system is a remarkably complex set of instructions; there are scheduled sequence jobs (i.e., user applications) or allocations of hardware resources such as the central processing unit, main memory, and peripheral subsystems.

Security Measures: The LAN channels of multiple architectures into a gateway convert data as it passes between systems. WANs connect various computer workstations and smaller networks to larger systems over greater geographical areas. The focus now is on preventing unauthorized access to important server and hardware components. Digital technology will now enforce the proposed security implementations emphasizing the systems three-tier dynamic Digital Network Architecture (DNA) framework following the Digital Equipment Corporation (DEC) model. (Appendix, Figure 11)

A network-of-network operating system is a remarkably complex set of instructions or allocations of hardware resources such as the central processing unit, main memory, and peripheral subsystems. Each username and password allows client users to access network files or to enter a personal gateway into an Intranet Web site. Security breaches occur if skilled intruders are allowed to crack passwords. Accessible business associations impose greater security risks to the organization. Data encryption utilities improve access security methods. With this new architecture, a systematic key encryption is mirrored seeing that passwords encrypt and decrypt data; although a public-key encryption is generated, one key encrypts data and the other decrypts.

Our original application software was developed in 1983 for use with an Informix/3B2 Telnet terminal system and eventually converted to UNIX ANSI C. The security issues were low and not a focus of the system design with no added security provisions. The previous systems UNIX commands are adapted into a new GUI shell. Telnet settings will be updated and are no longer necessary. The server's components are C++-based and optimized for integration with enterprise business solutions.

Network firewall security is enforcing protocols of network access jurisdiction. The firewall gatekeeper processes data to different sensing mechanisms, authenticating traffic and restricting network intruders according to access control policy. Intruders enter through the firewall tracer surveillance triggering sensors and algorithms powerful enough to follow a hacker's itinerary back to its origin. The firewall administrator stores access permissions and is taught to know specific priority restrictions. A firewall's footprint tracing tool is an important profile, logging and providing editing functions, although firewall protection strategy only protects against intrusion directly through the frame relay gate, which translates into inside intruders on the wrong side of the gate. Network administrators agree that outsourcing is the best method of deploying firewall protection because security experts prevent firewall contrition and attacks on the corporate network to evade system hackers.

Logic: The control unit directs operations according to instructions stored in block memory. UNIX and ANSI C program source code direct server processing. The control unit regulates data between memory buffers and the digital logic unit, routing data to output storage devices. Digital logical processor circuit registers temporarily store data from memory. Memory chip instructions remain inside the computer. Random-Access Memory (RAM) or internal main memory, receives the information and instructions from the microprocessor or input device.

Application agent software uses a combination of network-based rules, and client schedules integrating aging historical details automatically input from the application software into payroll and accounting system. Terminal network users access the subsystem of digital application hardware and software design that has evolved into a diverse multimedia environment. Authentication and permission protection is reconsidered in addition to a decision for something new implemented into the design architecture for preventing unauthorized access and enhanced security control protection. The successful log-on sequence interface is registered and stored on the SUN Enterprise 250-server in the standard COBRA infrastructure. After reviewing characteristic trade-offs and measuring alternatives, a new component is considered for the system.

Fuzzy Logic: Knowledge about fuzzy variable relationships is modeled by communicated keyword sets. An operator's job decisions are based on a hierarchy of precise Boolean logic identified objects or explicit events not originally identified. Object-defined sets also correspond to symbols or binaries. Fuzzy rules and logic combine with tables, and fuzzy decision control mechanism capture the inherent property of physical real-time variables. Human knowledge or behavior demonstrate a fuzzy relationship that is equivalent to input-output mapping or a transfer relationship.

Enhancing intelligent supervisory channels is a premise criticized in decades of machine coexistence. The computerized AI enhances speech recognition and security identification process, accumulates statistical data, and improves support service. Employees who understand and perform on traditional intelligence tasks require a stronger resource management discipline. Merging fuzzy AI by stimulating human and network channels improves coexistence with rapid-access machines. DNA technology demands characteristic ways of distributing knowledge, which sets its own objectives and would have us evaluate progress toward those objectives in terms of its own criteria and logic.

Digital Word Process: From an information security perspective, the UNIX command shell opens the door to misuse and corruption of the system information that could cause unexpected accidental or intentional catastrophic events. The reengineering objective required a more flexible application platform to adapt to changing needs and expanded regional markets, as well as the new rouge interface to distribute scalable portability to an open axis DNA distributed architecture. The new VPN design should add exciting state of the art intelligent technologies and provide security authenticity and integrity with the IP infrastructure configurations.

Digital describes a device that represents the values in a form of binary digits or bits. For example, a microphone translates sound waves into electrical current then the computer digitizes waves through conventional devices. A voice activator recognizes spoken commands and records a series of microinstructions as instructed by computer. The samples are stored, resulting in bit charge, creating a body of processed word information. A standard word length unit is 8, 16, 32, or 64 bits. Automated digital packet inputs expand the enormous capacity of output denominations. The RF interfaced stratum is connected through DSL, TCP/IP, RJ45, Fiber Optic, MIDI, or nominal D-sub 25 dimensions.

Multiple manufacture devices respond to motorized fader control on the fly, triggered by a sequenced real-time snapshot updating nuance controls. Virtual potentiometers' input channels motorize fader output routing layers, adding flexibility of a computer screen, mouse, and keyboard and delivering conventional machine-driven spectra. Traditional toggles assign seventy-two recordable real-time human performance assignments and balance fiber-optic perspective in a multichannel assembly. Waveform signals are now synthesized in a variety of interesting hi- and low-pass filters. Furthermore, eight auxiliary internal masters are flexible, open-end methodology. Every aspect of the frame accurate synchronization is precision-built upon layers of virtual channel bus subsystems.

The Digital Signal (DS) or service represents a common carrier digital transmission. Higher capacity channels are constructed by multiplexing the lower bandwidth channels together with some additional electronic signature that cannot be forged. DS signature confirms that a document or e-mail originated from authorized individuals and secures authenticity. Digital Signal Processing (DSP) assimilates the circuit used in high-speed data manipulation integrated into sound cards, modems, and videoconferencing hardware. RF communication applications manipulate image and data acquisition components. The preferred software provides a nuclear solution of correct switch data from cataloged historical trends.

Synchronization: A digital sample clock is consistent throughout the system as a master sync source distributes the word clock signal to slave devices. The internal word clock slaves to the external word clock machines. Some digital interfaces are self-clocking—AES/EBU, SPDIF—and do not require a separate word clock, although they function as slaves. A master-slave designation program guides the MIDI Machine Control (MMC) and the Society of Motion Picture and Television Engineers' (SMPTE) 24-bit word-clock synchronization—AES/EBU, TDIF, and ADAT optical exchange. Sequencing software transfers simultaneous data distribution through fiber-optic, XLR, and RCA connections.

A synchronized sample rate conversion of 44.1 KHz and 48 KHz depth determines bits connected to each sample: 16-bit or 24-bit. Higher bit resolution adds dynamic range that captures accurate representation of real-time system imaging. The operating system folders offer transport control, level setting, storage, and MIDI virtual sequence channeled design. Available real-time visual units are GUI, SMPTE, and BBT. Location looping recalls numbered instant access transport cues. The new DNA regulates distributed information processing. The virtual network machine is configured to mimic and store process a real-time digital automation engine.

Diagnostics: Another important aspect to a VPN approach is the program that tests computer hardware and peripheral devices for correct operation. Some faults are easy to find and the diagnostic program will diagnose the problem correctly. Several system faults are difficult to determine and diagnostic programs automatically correct errors each time. Some faults occur under specific circumstances rather than every time the memory location is tested. A simple set of instructions check the Basic

Input and Output System (BIOS) when the computer is first turned on. The tests results are stored and are known as the Power On Self-Test (POST).

If an error occurs, the computer stops and displays a useful message on the screen. Routine utilities store valuable internal settings or monitor external inputs. Automated input diagnostic software store and retrieve information on employees, customer preference, and product enhancements. The DSP visual tracking outflows add meaningful object-folders of recognition and are the interface methods for monitoring input and output definitions. Digital clipping LED meters represent a full +4-dBu scale, digital word signal, 0 dBFS max, and 10-15 dB range, above the nominal level headroom. GUI meters toggle shared DSP tools, automatically corresponding to precise system automated control; loop punch tolerance is three seconds.

Computerized Machine Management

Automated system diagnostics and programmed backups handle an extremely large capacity of stored translation data. Storage device like hard disks, micro digital tape, and DVD/CD writers assist in the information channel distribution, translating bit charges received from the microprocessor or memory into magnetic fields. Energized digital technology extends capacity beyond the fixed Small Computer System Interface (SCSI) or the Enhanced Integrated Drive Electronics (EIDE) drive capacity. Other important properties of computerized machine control programming are Video Graphics Array (VGA) terminals which displays millions of colors and calibrate postscript printer output, electrically charged with cylinders of patterns representing folder documents like text—HTML, AVI, DV, MPEG, JPEG, GIF, PSD and so forth.

The VPN application now provides compatible solutions for exchanging operator and business management data as the system collects volume data from a variety of service switches projecting historical visualized trends. Agents act upon queuing and heuristic algorithms based on the dynamic environment and provide a vigorous method to monitor daily activity. Service input applications target a diverse terminal production model of affective computerized system resources.

Dynamic Host Approach: The system is now based on several network interface cards configured to allocate IP addresses and other relationship information automatically. The dynamic data exchange is a technique used for communication applications running simultaneously which establish a two-way connection between requests and transmission data to each program alternatively.

For example, a web page specialist prepares audio, video, and animation on studio workstation computers. Computer Aided Design (CAD) programs assist with engineering and architecture planning. Workstations supply input for graphical Web site design that enhance the VPN characteristics and enable users to search, create, exchange, and display high resolution graphic pictures of processed catalog, or hear and display digital sound. Pictures created with specialized tools enhance web site scalability. The dynamic host transfers information from a spreadsheet to a word processor or database by pressing macro harmonious commands.

On important observation is the specified computer user cannot insert, change, or delete systems data without some human intervention or permission because the operating system responds to commands and coordinates the flow of information between the input and output sensing. The UNIX operating system serves as the dynamic host interface.

Channel Leverage: Automated technologies' uncertain future and increasing financial consequence hinder the growth potential of reliable business solutions. The implications of human resource management in the twenty-first century coexist in a civilization of automated machine technology motivated to deliver superior client services. Implementing automated technology offers no guarantee of actually enhancing the quality of products or company profits. Additionally, the increasing operational costs of implementing WAN-skilled channel-leveraging employees are troubling in the long term.

The secured distribution of digital information is establishing alternatives in a saturated Internet industry. Automated network solutions for today's virtual private channels assist managing information systems. A strategy of human resource management to deliver reliable network channels is balancing organizational growth in a strategic natural environment. According to numerous business consultants who also explored the enterprise venture, an organization is thriving through competitive human resource productivity motivated by customer demand. Machine-intelligent channeling of resource direction compared to statistical correlation measures help determine ad hoc training alternatives and is profitable and adaptable to collaborative influence.

IP Management: A unique 32-bit number identifies the computer on the network or some other Internet Protocol. Additionally, the IP address is configured in four numbers separated by dots and divided into two parts. The network address is a high-order of bits and the host address comprises the remaining. The host part of the address can be subdivided to allow for a subnet address. Computers tend to refer to domain names as a reasonable alternative. Shelf software functions rather than the expanded scope supported by later commercial IP address solutions. IP support is usually not easily available, which is unacceptable for most companies.

Licensing of software discourages reverse engineering, copying, or mass distribution of intellectual property. Specific commercial software quality increases license fees, which motivates a trusted IP design scheme to address computers by name. Host tables refer to machines by name instead of number LEDs. The IP address management scheme is actually the host table files IP address with names. The host table stratifies name mapping to the IP address, and reverse mapping of a host's IP address to search for a name. The host table is changed if a Network Interface Card (NIC) is added or a host deleted, altering the host's IP address, master host tables, and FTP. Commercial IP machine assembly simplifies address management, although with closed functionality.

Electronic Commerce: The e-commerce model is a standardized transaction exchange FTP expanding self-service production and innovative products. The increased security precaution, a realistic economic solution incentive, benefits large-scale participation. The explosion in e-commerce business awareness is developing into long-term client relationships, but the network environment, plagued with hype and jargon, is fueling uncertainty of client transactions. Common alliances improve the tandem operations of e-commerce security.

Market sensation is limited by collateral cost to merchants and participants. Channel subscriber links authenticate with unparalleled security so that safe electronic transactions can be conducted. Furthermore, firewall encryption precautions benefit consumers, making protected links available to reliable participating clients. Information is transferred for merchant processing, collecting key data fields, storing audit trails, and retaining tax records of system transactions. The firewall transaction center mirrors authentication, contributing part of a higher dimension protection layer and enhancing consumer channels. Viable associates appreciate e-commerce security. Experts agree that expanded business channeling security ensures transactions are private and remain safe.

Sharing and Trading: The role of the computer in supervisory control is compared to the human operator extending capabilities of human tasking activities. Extending and sharing control implies that humans and the computer are both aspects of the same system. Archiving is a replacement contingency of system authority. The human sharing or "exchanged" control mechanism is a paramount environmental design relationship, real-time assignment of task, or deviation error of a nominal condition. Human potential is a willfully mastered restraint cultivated through forms of comparative agent hierarchy definition.

For example, customer service managers recommend service satisfaction, so employees understand how much the company values its customers. The fundamental concern of is good training, motivation of users, and evaluation of client services. WAN administrative priorities alter model blueprints, determine performance expectation, and thus provide useful tools for competitive strategy. Economizing client's diverse relationships or personalities consequently increases Internet leveraging. Similarly, channeling commerce motivate peace of mind, enhanced security precautions, and has increased productivity by training employees to communicate with technology as a focused application to job function, utilizing invention for today's management information systems.

Uniqueness of the Method

Expert's consensus of the industry is that the corporate security system fails if a catastrophic breach results in damage to the organization. Furthermore, a restricted database is frequently vulnerable to intruders and. Databases actually stolen from the server contain not-yet-released information or products. As infrastructure technology expands into the twenty-first century, challenging methods of routine service create a lucrative environment for criminals. Although the new wave of criminal mischief causes turmoil within the industry, several high-speed providers distribute secure DSL information within a secured LAN distribution center.

A global paradox of security change is rapidly occurring initiated by the recent hijackings resulting in the bombing of the Pentagon and the World Trade Center. There are increased concerns for reliable WAN privacy and user footprint tracking for Internet security. W3C Internet provides a viable channeling method to link company associates and mobile constituents. Hackers and techno terrorists cause virtual damage to readily available e-commerce information that is in the billions of dollars.

Executives anticipate a disciplined safety margin of impenetrable infrastructure security integrity. A comparative analysis of several Internet technologies determined Internet firewall technologies are useful in virtually every aspect of today's managed information systems. Corporations might not fully realize exposure to unsafe situations. Network security is crucial to the continued success of the organization and necessitates protection of server integrity in a high-tech and competitive environment.

Market sensation is limited by collateral cost to merchants and participants. Channel subscriber links authentication with unparalleled security so that safe electronic transactions can be conducted. Furthermore, firewall encryption precautions benefit consumers, making protected links available to reliable participating clients. Information is transferred for merchant processing, collecting key data fields, storing audit trails, and retaining tax records of system transactions.

The described VPN solution is providing a security-conscious environment to maintain a safe e-commerce transaction scheme. The firewall transaction center mirrors authentication, contributing part of a higher dimension protection layer and enhancing consumer channels. Viable risk associates appreciate e-commerce security. Experts agree that expanded channeling security ensures transactions are private and remain safe. The e-commerce model is a standardized transaction protocol expanding self-service production and innovative products. The increased security precaution, a realistic economic solution incentive, benefits large-scale participation. (Appendix, Figure12)

Successful Resource Exchange Potential

The potential tools surveyed are microprocessors, memory chips, input devices, storage devices, and output devices. Read-Only Memory and Random-Access Memory (ROM-RAM) modules store microprocessor instruction. Backup resource servers are programmed and configured to communicate with notebook computers. Personal computer users do not usually connect to a hub or terminal routing gateway. Conventional software is connecting LAN to Internet frame cloud IP and the browser software accurately locates a URL that contains an IP address.

The year 2000 reinvented obsolete machine procedure, inspiring in-house training and repairs or system upgrades to outsourced venders providing no clear future direction to profit share of a growing Internet marketplace. The conservative approach is upgrading reliable technology and maintaining cash flow with a reasonable assurance of acceptable profit margin below the expectation of automation upgrade exposures. The regional growth is increasing long-term corporate forecast. Management rendered a plan of action to deal with the COBRA defined standards of network architecture aptitude and devise a control mechanism for government regulations. The immediate administrative remedies are focused on short—as well as long-term resource plans of realistic solutions.

The in-house development solution is experimenting with human resource training through meetings, where conflict resolutions are practical and a firewall gateway is protecting critical business information. Network certified employees insure the successful resource exchange continues and are trained and to protect password authenticity. Audio and visual digital production is designed to validate real-time multimedia urgency. Criteria include supports for human resource activities through mechanized task and preset machine control performance. Furthermore, the growing database of client contracts and profiles is actually a visualized plan for future electronic media distribution.

Computerization is defining mechanisms to mimic the flow control of information distribution. The illusion of virtual settings is recorded as control knobs are moved and embedded objects are grouped in meaningful arrays. Criteria include support for human resource activities through mechanized task and machine control performance of common alliances that improve the tandem operations of e-commerce security design. VPN enterprise workstations use DNA framework for e-process consulting issues and are optimized to visualize a plan for future electronic media. The use of graphics, sound, and computer-interface tools creates the intrusion-free gateway of production into successful rapid-access WAN technology. (Appendix, Figure 13)

Summary of Chapter 3

This chapter presented the description of the methodology used in the study. The previous reengineering application software functionality and its improved security procedures were introduced in addition to the new terminology related to that represented section. A description of the methods used to determine validity to the study is described. The C++ software and the VPN application reengineering requirements are discussed which include marketing strategy heuristics in the development of the database of study section, approach to customers business needs in the computerized procedure environment, the input-output process, as well as a new GUI system architecture, and its security concerns were presented in the VPN section of the study. Subsequently, the security consideration was discussed further to illustrate server and firewall integrity precautions that are now enforcing WAN security. The design and development trade-offs and COBRA interface definitions were also presented. Important concepts of the UNIX and Windows security configuration methods include channel routers, wireless mobility, IP address, and important DNA computer logic of computerization necessary to affectively operate the new VPN. The methods of synchronizing and diagnostic tools are discussed which relate to introduced computer applications. Relevance of software and approach to national business expansion is based on expert's opinions in the industry and review of literature results, with emphasis on considering information protection. The major UNIX user's security settings and their application interface configurations were discussed. The implementation of the VPN security considerations was presented along with the security exchange characteristics including key macro definitions. Uniqueness and limitation of the security and chosen methodology was also explained at the end of the chapter. Illustrations of the new COBRA three-tier architecture, original LAN architecture, DNA scheme, and improved cable-segment RF extension connections are referenced in the Appendix.

Chapter 4

Data Analysis

Overview

This chapter presents an analysis of the study that begins with an overview of the reengineering experience, including a detailed description of the security utilities and VPN design component functions. The investigation discusses the firewall and applications of system hardware and high-availability requirements of computerized competence. The TCP/IP issues recognize unexpected dependencies of implementing voice and data recognition model into intelligent supervisory control. Cisco and Annex network services integrate a process that shapes actions into the enterprise.

Reengineering software treatment of the user security utility functions explains application requirements. Furthermore, the details of the firewall security measures show how the UNIX interface library functions integrate into the utility register server. The system scrutiny also discusses issues and dependencies of implementing the security functions. A COBRA distributed middleware hosts a three-tier business component server. Finally, the chapter ends with the consequence of information privacy, protection, and factors that could limit data results.

Implementing the VPN

The problems of educating personnel present various difficulties associated with proper training of employees to use and communicate effective business information in a multiregional community and sometimes segregate company operations. Network alliance, established through secured telecommunication, promotes serious customer service dedicated to increased corporation growth and customer quality assurance.

In addition, compulsory credentials indicate verification of the required administrative trade-offs.

Standardized measures of efficiency frequently compete for the market share of their services. Each professional deserves recognition to appropriately store into a database or maintain information regarding supplemental services on an annual follow-up. Complaints determine priority rating, rapidly changing the plan and work that communicate information. The level of detail required to maintain effective network communications is a focused dedication to the technological advancements in the business community. For instance, modeling the linked machine connections for information distribution is a critical factor about formulating sound technology decisions, applying focus and enforcing administrative priorities.

A business system investigation does not necessarily include in-depth market analysis or web strategy of market evaluation research. Interactive machines trained to mimic human intelligence improve capabilities in pattern matching and methods to describe object events included in a knowledge base. Evaluated research does uncover obstacles in a deployment project operation and presents future quantity objectives of "Target Market."

The new paradigm of the marketplace is the "virtual community," a social or convergence entity created, in effect, by the proliferation of Information Technology (IT), or the emergence of a cross-disciplinary field of knowledge known as Information Science (IS), and the new reliance of organizations and institutions based upon novel computer-based electronic technologies and networks. Additionally, this analysis will examine the relationship between IT and the virtual communities that initiate individuals, organizations, as institutions to join the electronic convergence by means of stored, transmitted, retrieved, analyzed, and disseminated information.

Specifically, this analysis examines the critical issues that are affecting the shape and future direction of this field. Then concentrate on the commercial aspects of IT and IS. IS conducts research into the nature of RF carrier information, IT creation, organizational use, and model requirements that discuss interaction between people, information, or combination of conceptual structures with appropriate technology in the design of systems sharing, retrieval, and access, as well as information assimilation processing.

Given this broad and sweeping definition of IT, the rationale of this analysis is to identify the generally important or critical problems, issues, or concerns that are currently exerting influence over the field. IS knows few boundaries, overlaps many disparate fields, then draws upon their unique ideas or methods of discovered results. Included studies are in mathematics and statistics, computer science, Artificial Intelligence (AI), and human-computer interactions. Additionally, IS can assemble a coherent constituent of cognitive linguistics, with an epigrammatic philosophy of knowledge, communication, education, economics, political science, sociology, and administrative management procedure.

Varieties of disciplines intersect within IT and the inherent complexity. Disciplines influence selection of a limited number of major concerns to a somewhat difficult task. The significance in that, not-for-profit or profit-oriented successful ventures in the future gain competitive advantage, or allocate scarce resources to achieve maximum benefit, which capitalize upon emergent technologies. In the wake of this revolution, the VPN has emerged and is very quickly becoming an increasingly significant academic, as well as practical discipline.

Recently, the American Society for Information Science (ASIS) held a summit meeting in Boston and described a critical event in the development of the field of information architecture, which encompasses system design and development, information retrieval, graphic design, knowledge of organizational and user behavior, and skills in the area of existing and emerging technologies. As a relatively new discipline, IT has quickly generated a myriad of studies that attempt to identify the critical areas needing research, or refinement as the field advances.

Programming is the preparation and writing of detailed instructions for machine-driven activities. These instructions tell the computer what data to use and what sequence of operations to perform with the data. Computer scientists write custom sets of instructions emulating business operations by means of stylish workstations. A software compiler is required for executable conversions. Assemblers translate instructions and are a low-level machine language. Each program requires an interpreter to compile, assemble, and operate systems function as "smart programs" that enable a computer to carry out complicated instructions. The user communicates the code; smart program communicates output, then channels to central processing unit preparing a software program. A utility tool debugs—that is, tests for mistakes, on the real-world computer network platform.

Computers work directly with programs written in software compatible languages. Actually, the smart program first translates the written program into machine language. As computers have grown more powerful, the size of instructions has equally grown dramatically, increasing the central transport bus processing demand. Some modern applications contain tens of millions lines of programming instructions.

Virtual computerization has come to be the popular description of the twenty-first century organization. The principle is that time and space are no longer the organizing foundations that they once were. A fundamental principle of VPN organization does not exist in one place or perhaps even one time. IT exists whenever, and wherever, the purchase goods happen to be. Today's company may actually have virtual organizations in-house, just as they have outlander communities of practice. General points of interest include controlling network technology, user monitoring, dynamic configuration management, and client quality control. Essential for systems is a disaster relief and recovery consideration to inspire a survey of high availability design techniques and offer continued commerce operations in the event of an emergency or system crash. Diagnostic assessment recommended backup procedures and state mandatory rules of procedural options. Utilities running smoothly preserve executive desires for a secure management environment.

System failure will have a visible impact on network operation and financial obligations. Furthermore, monitoring is a critical contributor to firewall security scheme and reliable delivery of quality controlled independent system functions. In addition, technology-based products affect the course of business discussions. The important question is, will "changes in technology" directly affect present operation of business information and threaten employees, or add significant nobility to their routine machine operations? Moreover, consider pending legal issues before the U.S. Supreme Court of patents or trademarks that affect a low-technology product whose manufacture or distribution triumph involves the vending of patented watermark processes.

High-technology products and services directly affect business in many ways. In distribution, for example, there might be a change in scanning technology, retail point-of-sale systems, or even video displays. In other business channels, technology changes the nature of the goods produced such as surfing the Internet, mobility of cell phones, or high-density video transmission that did not exist a few years ago. Does VPN technology affect advertising, sales, or the Internet marketing strategy? Can a Web site change the mode to conduct business? For example, the same competitive agency that depends on a computerized ordering might discover competition in subcontracting of direct Internet ordering.

Not all systems depend on computerization and executives might discover irrelevant business operations. Market segmentation provides the foundation to demonstrate the reasonable balances.

UNIX / WINDOWS 2000 / XP network capabilities include:

1. Complete e-mail facilities on the Internet, CompuServe, America-Online, and Bellsouth DSL, T1 connections, for working with clients directly through e-mail delivery of information.
2. Provide presentation facilities for preparation and delivery of multimedia presentations on Windows 2000 / XP machines in formats including on-disk, live presentation, client support, or audio-video graphics array.
3. Develop adequate desktop publishing facilities for delivery of regular retainer reports; envisage output analysis and market research description.
4. Provide 24-hour TCP/IP Annex communication, monitor corporate tracking information, payable and receivable database, decision support from infrastructure, recording and security surveillance.
5. Configured Cisco 3600 series routers authenticate through the firewall gateway of virus protection and encryption security.

Technology deployment frequently begins with a negative cash flow factor. This analysis considers only a limited number of the critical concerns facing the discipline and there are additional issues that may be significant to the problems discussed herein. For example, an IT advancement in record time creates congruence and synergy between various subspecialties, although is a challenging task. What expanded is "implicit culture," inherently complex in data, and information-rich, burdened with numerous uncertainties. Limited research is available to evaluate the myriad concerns generated by the virtual phenomenon, which tend to infringe on individual rights and, to a lesser degree, specific organizations of "stratified members." Such research, inevitably fragmented and diverse, consequently identifies significant issues related to IT professionals, relatively in early stages of discipline development.

Commerce tools manage content, as effective customer data that is not enough to determine key factors of relationships that gain advantage of the texture, feel, and terms to process abstract design. The nature of information and degree of the personal exchange establish contacts, community partners, and a collaboration of single or multithread solutions. Personal resources address the needs, risks, or possibly negotiations to dynamic co-creations. The e-process drives the enormous captivity of channel sourcing that is characteristic of individualized reporting.

Component Analysis

The methodology employed herein was a qualitative, descriptive research effort. An extensive review of relevant literature, spurred in part by texts read in the MIS curriculum or the analysis, facilitated the identification of the critical issues that are now affecting IT in real-world and academic environments. Qualitative or descriptive research is extremely useful in establishing the nature of a research problem that leads to empirical analysis.

Programmers use C++ instruction symbols and various everyday expressions, such as "READ," "PRINT," and "STOP." For example, if a task involves processing business data, the programmer would most likely use Common Business-Oriented Language (COBOL). Programming a computer to solve complicated scientific problems might require the use of a mathematically-oriented language, such as FORTRAN (FORmula TRANslation). Object-Oriented Programming (OOP) combines entities to create a whole sequence program. OOP relieves programmers of the need to recreate sections of code in long programs. Embedded systems designers create applications that use existing components and minimize the number of software elements.

Commanding in this technology is the Microsoft Windows family of products expanding the number of choices for developers of embedded solutions. A variety of networking products provides a set of 32-bit operating tools in the company of numerous unique advantages not found in the supplementary embedded operation systems. Members have the ability to expand existing, off-the-shelf hardware and software components. Embedded operating structure provides analytic support for a broad range of development tools and resources used by the developer community. In addition, the Windows descendant products support a wide range of commercial applications including data distributed messaging and web-based advantages.

The traditional telephone boundaries are extending data networking into an emerging market that values the commercial advantage of multimedia-capable PCs. Furthermore, lowering cost of communications, combined with application tracking notes that provide user assistance, formulate the reality for network managers to transport real-time voice conversation over RF carrier transmissions. Bandwidth augmented transmission of voice protocol standards are expanding management tools that determine if an existing network can manage information over IP, ATM, and frame relay communications.

Voice and Data Internetworking: Each top quality telecommunication network reviewed implemented voice transmission communications. The value is flexibility to use a computer screen, mouse, and keyboard, in addition to conventional voice and data spectra. Digital waveforms define user auxiliary channel features that include bouncing signal perspectives, then storing routine snapshots. Line level motorized fader movement is memorizing on the fly, guided by the Midi Machine Control (MMC) timing reference. System IP also defines and configures network specific capturing devices.

Toggle switches, in conjunction to moving with numerous assignments on the display, provide visualized waveform signals. Real-time efficiency organizes network architecture. Media such as wire, fiber-optic cable, or RF carrier waves transfer data connecting nodes to secured users, establishing support services, training, strategy, and motivation tools. Administrators present different perspectives and improper network protocols create significant problems of transfer and compatibility. Debugging the software or connections insure accurate transfer positioning, data movement, and compatible interpretation through printers and video of discrete measurements. Hardware and software are compatible with voice and data technologies.

Enterprise Consulting: Technology utilizes administered management for evaluation of service that represents many expectations to a client, as the company becomes an extension of its business development channels replacing routine functions. This begins with complete understanding of the apparent situation, objectives, and constraints. Company focus is representing the client confidentially, sifting through contemporary promotion opportunities, which represent the possible allies, vendors, and registered layer channels.

With the available hardware and software choices today, useful solutions to manage today's enterprise tasking activities increase initial expenditures that reflect the uncertainties of on time payment, as well as the government regulates a community. Each business contract alters cash flow and certainly the client's ability to pay. However, captive control spending on a per-project agreement is a useful perspective, as beneficial consulting differs for each consumer. To identify trends, then obtain a capital improvement, is a process of enlarging a lucrative organization, then following proven basis for costs, in addition to persuasive arguments for future possibilities. Increasing market and sales opportunities through the Internal Support System (ISS) integrate graphical Internet presence.

Important dimensions of enterprise management require integration or "bridge-building" for the future. The first dimension involves the many disciplines that constitute IT, which have been more or less independent of one another, essential to infrastructure science.

The very newness of the field has created fragmentation among internal disciplines and mandated a new understanding of the manifold specialties and subspecialties that together create IT. These include such varied specialties as Artificial Intelligence (AI), communications, and theory of cyber metrics, cybernetics, then as consequence liabilities, linguistic subspecialties of phonetics, semantics, semiotics, speech recognition, telecommunications, systemic, and neural system science emerged.

Each discipline focus is on the phenomenon of information distribution as the object of a query, of a universal practice between distinct stratified cultures, which function in a comparative isolation. Achievements of goals necessitate the full multistep process of definition, tool selection, or resource implementation that results in benefits. Specialties have their own languages as entities using knowledgeable words. Consequently, the lexicon provides integration of each discipline. IT no longer handles synthetic data or just manages information systems, but concentrates on unquantifiable, yet valuable resources of knowledge.

The second dimension identified requires further work with the three building blocks of, people, knowledge, and tools. Coordination of disparate tools, disciplines, or knowledge bases and professionals, are necessary stakeholders in the new era VPN. Organizations such as the Application Specific Integrated System (ASIS) dynamics should play a significant role as a multidisciplinary forum, which can stimulate training of the information infrastructure disciplines. Discussing the representation will augment current speech-based retrieval of interface systems. Further integration of the disciplines illustrates an emergent instrument that crosses a broad spectrum of varied treatment.

The third dimension deals with the key areas of enterprise discipline: education, research, and practice. Higher education, particularly graduate education, is an ideal focus leading to greater integration of subsumed practice. Research tools that design, develop, and improve the routine relationships between people and knowledge, or creation, capture, storage, preservation of identification, dissemination, and operation. Social implications suggest the field is beyond data management within organizations and influences the interpersonal communications. Presence also is a significant challenge to business and social applications. With confluence activities, convergence in media as visual, print, audio, and multimedia challenge management to rethink the way practice occurs.

IT specialists recognize that information discipline management demands enthusiasm in assessing the quality, examining the utility, evaluating the impact, then measure the influence of information effectiveness of an organization. Noting the effects of information distribution on an organization's financial position or more significantly, enterprise, introduces vitally important ethical considerations, held by associations, IT professionals, and shareholders.

Monitoring e-mails or reporting sessions are comprehensive although to track everything on a daily basis is limited, cumbersome, and enlists administrators to monitor a proximity solution. Enterprise security can intercept viruses, hackers, or crackers and challenge competitors with the ability to respond to a breach, protecting vital credit card information, destruction of personal files, or prevent consumption of transparent vulnerabilities.

Managers are formulating a strategy and implementing policy within the context of ethical uncertainty. In simplistic instances, ethically questionable outcomes are a direct result of principled intentions. Complex instances of unethical outcomes result from practiced intentions that are themselves based on sensible business and managerial principles. Given the dynamics of the problems, possibilities of disseminating confidential information compound the IT professional enforcement of ethical codes and policy.

WAN Interface Cards: In computing, hardware device connection allows computer telecommunication to a computer network. SCSI is a bus channel to carry data between the computer and other devices (e.g., printer or scanner). Alternatively, the chosen interface bridges between a series of devices. Although cable connections consist of optical fiber components, collections of communications, and node pathways, the actual size of a server bus adjusts proportion to the number of channeled connections. Bus slots inside the computers allow communication devices, controller cards, or host adapter circuit boards to operate. In addition, interface cards, such as video cards and elaborate audio processing cards operate programs simultaneously allowing continuous integration of common parallel standards for fixed or hot swappable drives.

The control electronics reside on the SCSI drive, not the interface card. The low cost of Integrated Device Electronics (IDE) for the mass market is appropriate for Advanced Technology Attachment (ATA). Enhanced IDE is an updated version of IDE that improves on speed and adds support for drives larger than 528 MB. ATA Packet Interface (ATAPI) is a software protocol that allows support for CD-ROM drives on an IDE/EIDE interface.

Minicomputer interfaces of priority are not intelligent and specifically designed for one device. A special interface is required for each node of a fixed disk. Prior to SCSI, minicomputer users changed both software and hardware to support new devices. SCSI cards provide even greater flexibility and improvement of data transfer at 20 MB/second on a 16-bit bus. Innovative interface markets are competing to provide the link between the high-speeds of SCSI-3 and SCSI-4 serial devices, fiber-optic channels, and Apple's Firewire (i.e., IEEE-1394) that provides a transfer rate 400-1000 MB/second.

TCP/IP: The protocol used by the Internet is still the technology that fuels networking services like the World Wide Web, Internet Relay Chat, and electronic mail abbreviations for a set of network protocols, developed principally by the U.S. Department of Defense. TCP/IP has always been the principal group protocol utilized by UNIX and supports competing operating systems.

The modem is a contraction of modulator/demodulator or device for transmitting computer data over telephone lines. The modem converts the digital signals to analog, then back to digital again. The VPN global network and Java provide portable program access and XML enables programmers to create PXML and XHTML 1.0, the latest version of HTML from the W3 Consortium. XML also provides the last piece of technology required to develop truly portable applications. High-speed standards emerge as suitable types of facts, including voice and video transfer valuable resources. Each nerve cell, in the middle, and bottom layers, receives input from several database sources only when the amount of input exceeds a critical level output signals. The output process will include pattern recognition, machine learning, time-series forecasting, machine vision, and intelligent control or ability to learn. To recognize the Internet layer is a responsibility of routing packets, checking IP address, subnet mask, and default gateway. (Appendix, Table 4)

Configuration problems occur with an incorrect routing table. Table errors may send packets of information somewhere between systems, attempting invalid communication. Without applicable connections, communication is not possible. Problems occur if the parameters have a configuration error or incorrect IP address, which demands broadcast quality streaming, videoconferencing, distance learning, surveillance, and Intranet training applications. Actually, MPEG-2 is producing impressive results over a gigabyte VPN, in today's leading edge envelope technologies. Measurable objectives improved the IP production of the market-share objectives. (Appendix, Table 5)

Alternative Approach Issues

The approach control factors consist of executing system start-up and shutdown, adding and deleting users, protecting user data from modification, either by users or through hardware failure, managing and adding peripheral devices such as disks, printers, etc., maintaining WAN connections between subsystems, ensuring the logic runs smoothly and provides necessary client-services.

UNIX operating systems have a history of about thirty years. Macintosh's fifteen-year history continues to rival Microsoft Windows operating scheme, although is particularly shallow for operating systems, in terms of network administrative services. The advantage of living in the modern age is, computers are faster and offer tracking assistance, with lower cost memory and greater efficiency. Future generations will laugh at the sluggish velocity of present-day technology. UNIX is among others who developed useful program code for today's information systems. VPN environments build from a group of What You See Is What You Get (WYSIWYG) software. The deployment theory is to run small useful programs so they fit into configured memory in various combinations of output results.

Open systems are similar to a client-server architecture, although it implies the code phrase. UNIX cooperation systems allow flexibility, which most vendors cannot submit. Proprietary solutions of switching merchants are possible if problems arise, although still a difficult issue to resolve. Multiple platform features allow concurrent administrative services that challenges administering the operating system. The protocols determine a valued administration procedure. The philosophy is the "toolbox approach" to a multitude of commands and registered components. The initial focus is providing quality, open-end networking as a vital tool for achieving commerce sensation. The UNIX flexibility collaborations lead to an authentic practice.

Process is as fundamental to UNIX systems as breathing, although represents an instance of an executing program. Several procedures appear to be running simultaneously and technologies' management role is data manipulation and user tracking. Experience and programming skills are required to model the implement procedure. In addition, process control of subsystems is required scheduling, memory management, and communications stored in a table that contains an entry for each running service, in several nominal states. To recognize a suitable approach the system administrator understands various states of a particular process hindrance and is familiar with command Syntax. (Appendix, Table 6)

The fact is that government has become increasingly involved in setting IT policies and consequently models a scheme of practiced VPN standards. Conviction and debates will eventually determine whether government or the market should drive IT professionals and their enterprise policy. Infrastructure architecture serves the broad public as well as commercial, government, and industrial interests. Legitimate fears exist of big money dominating the global marketplace, then minimizing the ability of the ordinary citizen to leverage the social, intellectual, commercial, and political benefits of worldwide technologies. The link to the political, social, economic, and ethical demands laminate programming paradigms of applications that reflect the significant stratum.

Computerized task procedures are confronting IT professionals, regardless of their particular academic or professional field of specialization, capturing human intelligence in hardware and software applications. The routine analysis is to facilitate a type of asynchronous-based system that shapes the infrastructure productivity. Understanding the underlying mathematics of Boolean functions involve any number of variables whose values and judgments are binary. The "consumers" of modern systems integrate human intelligence into useful schemes that are artificial and the control of a critical task, which IT professionals must address.

Software integrates as part of the enabling technologies of the Web. What fails to transpire is a broadly applicable program language and universal hardware platform, which allows members to easily access one database. In this context, the commonplace of computerization has yet to revolutionize the world of information distribution and is now approaching sequenced sharing as accessible and useful. Resource management alternatives are a balancing of a vitally important bridge between the technocrat and the organization. Technically competent and proficient designers model adequate awareness of how the system performs tracking ordinary human performers and maintain the equilibrium influence.

The UNIX methodology altered the network revolution and proved as an effective communication medium today. Groups are attempting to alter dynamics that create appropriate standards for use by IT professionals, working to create information products or services. Liability processes recognize part of the value of relationships that drain valuable resources. Standards vary as diverse groups have pursued dissimilar goals that extend hardware and software systems for specialization. IT professionals are becoming increasingly aware of issues and must interact effectively with the dedicated infrastructure externally, which is extremely critical to cultivate effective standards that become an industry-wide policy; firms and institutions dedicate to revealing new alternatives.

Creativity, a characteristic of the media revolution, appreciates valuable standards essential for cross-application coordination. Fears of monopolies on specific hardware and software platforms speculate future political concerns, or the difficulty of moving readily from one hardware platform to another, or from software solutions. Elements of the underlying concerns hinder achievement to full potentials. One of the goals of IT is to reduce such interface difficulties and make the interaction between humans and expanded systems more easily achieved. This is a critical issue because VPN organizations are hosts to a wide range of alternatives and often scrutinize very diverse databases and software applications. IT professionals certainly work with environments to function effectively and serve their constituents.

Related to this distribution is the importance of IT education. Network professionals in business or institutions must work collaboratively to ensure that new add hoc solutions prepare for real-world discipline practice. For about one month, chief executive officers (CEOs) study the "best practices" based on research and teaching collaborations. IT professionals are educated and trained for Intellectual property issues and the only catalyst of this effort is identifying the research needs of the cost restraints of subsidiary disciplines.

Automation in Supervisory Control

Distributed computing is the sharing of computing tasks among connected computers. There are several forms of distributed computing. A computer network is a group of computers interconnected to share information or resources. This is especially true for object channel distributions discussed throughout dissertation and is a top priority to administrative requirements.

Dispersal computing makes each worker an independent individual workstation equipped to receive input and process information locally without the aid of another computer, although only system administrators have permissions to install devices or alter configurations. Multiple terminals process independent tasks until compiled at each channel IP address locations. Distributed computing is a major step forward in supervisory control establishing resources to allow complex computing activities on smaller, less expensive machines. Collaborative supervisory computing requires a policy of distribution rules, sharing processed power, server resources, and software, which process Structure Query Language (SQL) requests. Distributed computerization improved corporate growth in regional satellite office linkages. (Appendix, Figure 14)

The strategy behind the VPN segmentation is the definitive choice of targeted market expenditures. Successful advancement is a combination of what makes members interesting, or convey characteristics of growth that regulate the goal's effort. Leveraging of internal and external sources facilitate standardization of IT activities in the laboratory and workplace settings. Similarly, educational processes tend to incorporate critical technology components that recognize that a professional supervisor control mechanism is a critical factor of monitoring professional activity. Exploring and mastering the complexities of IT within the context of science is well underway, although affective educational preparation segregates the future course in this field.

Real-world knowledge is establishing practice, comprehending the distributed information that entails an understanding of the cognitive and intellectual processes the user formulates and expresses as query and assesses information. On the National scale, professionals have an increasingly responsibility elevated to understanding of system design, hardware and software applications, operating platforms, and regulated user interfaces The three-year analyses of urban activity reveal a 50 percent increase in contractors after subsequently deploying supervisory standards in 1999. The receivables increased specifically due to expended electronic transfer internet relationships. (Appendix, Figure 15)

An additional concern that supervisory professionals must resolve is economic in nature. The VPN revolution led to an enormous financial investment of human and technological capital. In addition, effective control is an investment that must link to productivity and profitability. There have been a number of attempts to assess the impacts of the virtual organizational performances, although these studies tend to yield conflicting results. Researchers have been unable to conclude that increased spending results in enhanced key performance indicators.

IT conventional wisdom demands institutions to make substantial investments by upswing machine technology. However, the model is without a systematic plan to purge supervisory resources into the organization's overall business strategy. The views of a supervisory policy express a rigorous business plan of strategy. This is as true in the case of the nonprofit or institutional members, as well as for-profit businesses and industries. Offering coverage of the entire network spectrum presents problems and issues that allow full-scale release countermeasures. The supervisory approach introduces dependencies of formal activities, software trigger patches, and RPC settings. The supervisory control details user security utility functions that provide a feasible tracing protection. In addition, it accommodates the enriching professionals who control-managed resources of the model (e.g., schools, healthcare facilities, etc.).

Cisco Network Modules: Nodes detect idle state to harness seldom-used CPU power. Processing data through predefined channel relationships is typically at the top of the system hierarchy, solving integrity problems, and the inability of processing to represent logical data relationships easily. The Cisco 2600 series setup commands help configure the router prompts for information required to start functioning quickly and install series routers that includes up to six slots in which you can install modules. Two small slots Wo and W1 are available for Windows interface cards. Serial WAN cards install into either slot. The router numbers will identify interface modules. Numbers begin at zero for each interface type, then continue from right to left, and from top to bottom. Modules and LAN interface cards define interface slot number. A slot number in a unit identifies each individual NIC on a Cisco series router. Voice interfaces and unit numbering include the Interface type, chassis slot / voice module slot / voice interface (e.g., Fast Ethernet 0/0).

Configuration of the interface is specifying IP routing. Custom configurations depend on requirements of the routing protocols and common commands. After configuration, check the interface version controllers and protocols with the ping command echoing a request to a specified IP address returned as exclamation point.

Voice interfaces number differently from mode and interfaces described.

- Slot 0, Ethernet interface 0, referred to as Ethernet 0/0
- Slot 0, Ethernet interface 1, referred to as Ethernet 0/1
- Slot 0, Serial interface 0, referred to as Serial 0/0
- Slot 0, Serial interface 1, referred to as Serial 0/1
- Slot 1, Ethernet interface 0, referred to as Ethernet 1/0
- Slot 1, Ethernet interface 1, referred to as Ethernet 1/1
- Slot 1, Serial interface 0, referred to as Serial 1/0
- Slot 1, BRI 0; interface 1, referred to as BRI 1/0

Voice unit-numbering interface is as follows:

- Slot 1, voice module slot 0, voice interface 0, voice 1/0/0
- (closest to chassis slot 0)
- Slot 1, voice module slot 0, voice interviews 1, voice 1/0/1
- Slot 1, voice module slot 1, voice interface 0, voice 1/1/0
- Slot 1, voice module slot 1, voice interface 1, voice 1/1/1 (farthest from chassis slot 0)

The Cisco setup command displays the PC terminal emulation configuration. The process requirements also depend on network module or interface cards installed. Available for configuring are the fast interface parameters: a token ring, serial ports, frame relay, LAPB, X.25, ATM-DXI, SMDS, PPP encapsulations, asynchronous/synchronous interface, ISDN-BRI, ISDN service profile identifiers, E1,T1, ISDN, PRI mode configurations which include circuits or on a dedicated service line.

The Cisco channel nodes divide into modes as each command permits numerous configurations that differ on the router at any moment and dictate which mode is in effect. Entering a question mark (?) at the prompt offers a list of commands for each mode. The multiplex trunk configuration software interfaces to the master router or network module to gain greatest authority and flexibility. Routers prompt for the number of channels that the network module can support, which also depends on the level of the Codec complexity.

To integrate into an existing IP network, the initial configuration procedure depends on the topology and can only support only one type of high or medium compression complexity. In addition, the T1 support includes an integrated Data Service Unit / Channel Service Unit (DSU/CSU), configured at 1.544 Mbps or fractional T1 service. The enhanced Cisco commands provide the greatest power and flexibility.

Voice over IP (VoIP) enables the router to carry real-time voice traffic to enable the following benefits:

- Toll bypass
- Remote PBX presence over WAN
- Unified voice / data trunk
- Connect Plain Old Telephone Service (POTS) or Internet gateways
- Configuration of VoIP over an IP network
- Install voice network modules into the router dial plan
- Establish a telephony network based on companies dial plan
- Integrate dial plan telephony into existing IP network topology

Virtual Network Issues: The Commerce Department characterized the VPN situation as a digital segregate with significant social implications for the nation and people. Presidential candidates pledge to fight the digital divide connecting public schools, libraries, and government service centers to the Internet. Government pursues a policy towards financial assistance to achieve this long-term goal or possibly work to secure lower costs or reduced taxes for low-income Internet users. However, the distribution of information technologies in American access technologies is stratified along ethnic lines and boundaries created by wealth, race, geography, and technological expertise.

The Federal Communications Commission (FCC) works with the telecommunications industry to create linkages at lower costs for rural, underserved, or expensively served, as their urban counterparts. The FCC is also considering ways to spur private sectors' willingness to bear some of the financial burden for improving access among the lowest-income segments of the population. However, these matters are unresolved. Educational districts access or purchase expensive information technology, as communications infrastructures provide students with tools. Not all Americans share advantages equally and the potential to polarize a society already divided by such factors as race, ethnicity, and economic inequities.

American capitalists have been talented and audacious in exploiting the economic possibilities of technologies. The success of twentieth century technology is providing Americans with convenience, comfort, speed, hygiene, and abundance so obvious there is only trivial reason to look for other sources of fulfillment. Furthermore, the devaluation of traditional beliefs took on the exaggerated significance that served to advocate technocracy into technology. The information era incentives are increasingly capital intensive, leading to further disparities in access to such technologies and consequently, endorse a state-of-the-art instruction.

The use of technology shapes structure by establishing the ends that technologies will pursue or facilitate. In other words, the technology introduces potentials; these potentials shape applications of expertise. Similarly, technologies annex important knowledge. More importantly, it redefines the boundaries of liberties or freedom, intelligence, truth, fact, wisdom, or memory, and challenges history to recall nuances of token words. In addition, it is becoming a manageable entity, working to achieve specific goals and program objectives that are integral to the administrative definitions of values. The structure created is a private society or possibly a massive public sector of possession, in pursuit of cherished values and civil meanings. Extensive debates on issues continue over the valuation of e-commerce, which concern management and financial business circles.

Annex Network Administrator: The high-level prestige specialty of VPN adds administrative control security to manageable business operations. Through virtual technology, system managers can secure business relationships through a smart connection of encryption, utilizing the Intranet firewall within the Annex channel router and terminal server.

Although many capabilities and functions are available to system members, keeping information confidential presents problems to administrative permissions on system. Building a take possession of authentication encryption ensures accurate and original data entry. Telecommunication channel tunneling standards of network protocol prevent system intrusion, inherent for Internet security, offering solutions for interfaced applications, also addressing each primary business objective.

Additional Remote Access Service (RAS) connects outlanders, increasing the mobile work force, encouraging working at home, allowing expanded hours of service while inspiring valuable production incentives for corporate development in future VPN technologies. Company decision makers enjoy the enhanced ability to control system security using the automatic motivation technology. Annex services have become the basis for linkage analysis practice.

Annex technology builds secure relationships through authorization and encryption and a high-level maintenance solution that delivers Internet support service. The increased virtual presence of strategy analysis is improving customer relations. System administrators cannot overlook inherent weakness of the Annex management structure when implementing new technologies. A company's system operations follow users, as necessary, monitoring responsibilities which associates rely upon, recognize error, and question integrity and reliability of systems information.

Users want to communicate with the host network with the input syntax character or word that separates the options available for a parameter. Descriptions of what the administrators expect to install on the index of network installations include configurations, diagnostics, and troubleshooting. Annex communications provides easier access to serial devices including computers, modems, and printers connected through the LAN. The host becomes transparent as initial terminal behaves using the CLI argument commands. Additionally, they connect services through a set of channel connections of controllers, hubs, routers, switches, and power support. The representation of the dotted decimal notation facilitates a host connection. Annex CLI terminals connect to a serial-line port administrator and create a log-on session. (Appendix, Table 9)

Annex IP protocol or host name consists of four 8-bit decimal values from zero to 255 separated by periods (e.g., 192.168.254.5). The network administrators secure settings of speed, parity, number of data bits, etc., regulated from the serial port connected to the terminal server. This methodology sets a username and password based on the Annex message tracking authorization, followed by the password prompt, then checks authorization, entering the input password that does not echo.

After verification, the Annex replies with a permission granted followed by the number of times a user is to reenter typing errors, set on the security configuration. Permissions protect authorized intrusion from the Personal Identification Number (PIN) command line connection. The host view of the table displays names and addresses of on the Annex members. In addition, configuring as a port server connects printers, modems, or devices to the LAN. A rotary port selection range or groups of ports on the Telnet session facilitate connections to the Annex host address of the source connection. Arguments assist the CLI displaying a short description of command is Syntax. Detailed processors, memory, and network adapters address hardware. CLI commands are available on the current port with a summary of host names and addresses. Annexes listed on the host table display status information obtained from a host broadcast and utilize information on main servers. (Appendix, Table 10)

Security Valuation: The components carefully group together to simulate routine business operations and systematically record important strategy and security information for analysis. The frame relay is a standard for packet switching protocol and can run up to speeds of 2MB per second, providing rapid data transfer. Each channel throughput is passing data into the firewall (i.e., SMTP, FTP, HTTP, DNS, POP3, Restricted Telnet, Pass Out, and Translate All). Server machine-driven operations determine destination of transmission, convincing a motion of data.

Add-in-boards expand compatibility through higher bandwidth transfer speeds. The bus mastering is COBRA three-tier architecture except for older Industry Standard Architecture (ISA). The channel routing carries information from a sending to a receiving device as a physical medium within specific bandwidths. Generalization divides the bandwidth of the communications circuit into smaller packet increments. Point-to-point circuits provide twenty-four channels of 64 KB per second transfer of 1.544 MB per second Digital Subscriber Line (DSL). A statistical multiplexer divides the 1.544 MB per second bandwidth into 2464 kilobytes of voice data spectrums of annex control authentications.

Scalability of large storage capacity is available over the Firewall IP network. Multiple time base standards and conversions are in progress to ensure that backup archiving exists in subsystem architecture. Digital tape is the easiest storage medium to use efficiently that enables the next dimension of secure visual intelligence. Backup timed solutions are uninterrupted standards of overall management security objectives, balancing the goal of gateway standards

Delivery of asynchronous information is output to Cisco routing terminals for interpretation (i.e., printers, video display units, mixers, recorders, etc.). This certainly improves the TCP/IP performance communicating on the e-commerce channels. Application software layers of client components configure with the standard Windows 2000 / XP elements, Sun Enterprise server, and a COBRA distributed middleware client and server.

Components at the register server provides user authentication as applications operate the routine functions. Channels attempt to mimic the activity of the brain as the individual computers undertake specific tasks and contribute to the desired outcome. The primary component of the security application is a navigator browser that locates the register server and then queries IP connection port. Once located, the navigator prompts for a username and password set on the host machine.

The security applications IP addresses communicate between clients and servers through secure socket layers of communications across the WAN components. This procedure guarantees that the clients and servers authenticate to prevent digital clone using the configurable 40- or 128-bit RC4 encryption.

The browser calls for the username and password and then sends the information to the register server for authentication, calling a register log-on, which validate security input. The log-on interface request from the navigator register automatically calls the corresponding variations of the security utility functions. The navigator registers rely on the user security settings, corresponding to variation of the security utility functions that validate a new password against the password criteria and then carry out the actual process to amend. In addition, the user password on the host machine is acting like a gatekeeper that protects valuable business data from unauthorized access and protects import system information from possible misuse and abuse under the shell commands. Information and communication technology found a fertile ground in America by establishing persistent values in newness and security improvements.

Machine Technology as a Commerce Tool

Electronic exchange accounts for over a billion dollars each year in revenues. The phenomenal growth of e-commerce offers a unique opportunity for consumers, businesses, and sellers to "publicize" in a virtual marketplace and "capitalize" upon competition, generally exchanged to both parties of the potential business deal. Companies set up electronic markets searching for ways to maximize future profits through efficiency and competitiveness. A computerized market characterize the rapid exchange of inputs and outputs, with additional ability to create a nearly perfect competitive environment to foster economic equilibrium and stability.

Trends in e-commerce potential tend to reduce excessive supply and production by carefully identifying demands. The Internet landscape argues linking buyers and sellers eliminate transaction costs of intermediate staff eliminating inefficiencies. The competence adopting business models and strategies fundamentally differentiate products and services. With no boundaries or barriers to access the infrastructure, the Internet invites comparison shopping in both the consumer and business-to-business markets. Furthermore, the paradigm impacts business practices as traditional forms of sustainable advantage, as novel discoveries emerge.

Several principals characterize the influence of e-commerce initiative and reveal emerging business principals:

- Accessibility via the Internet increases market reach that in turn challenges traditional distribution channels and redefines the role of intermediaries.
- Long-term winners must master not only the distribution and service opportunities of the Internet, but also methods to accelerate and simplify transactions and related services through integrated automation and process of reengineering.
- E-commerce business models define a company's competitive advantage in much broader terms than just products.
- Wholesale introduces increased competition and markets, resulting in greater pressure to deliver better service and lower price, increasing the value of high quality, customized personal service.

Efficiency, carefully designed to target business strategies in diverse areas as marketing, sales, transaction, and service objectives, materialize interactive agent searches on behalf of their users or clients, fostering efficiency in e-commerce. Agents in the form of functions, replicate organizational utilities acting as intermediaries between the purchaser and a seller. Agents lower search costs to consumers and suppliers, thus reducing business deal costs that are associated with market inefficiencies. The growth in the Internet media suggests that consumer, business-to-business, and projected e-commerce channels are improved.

Shaping Intelligent Actions: A coalition of the system requires evaluation in:

- Ease of training motivation communications.
- Scalability of operations associated with upgrades.
- Goals, focused on administration's commitment to building process, national politics, and the education of user.
- Support through administrative funding.
- The timely resolution of issues creates a model of system readiness.

The business-to-business side of online transaction leads to better-informed purchasing decisions and enhanced tools (e.g., electronic data exchange). Companies using traditional tools lead to comparatively modest gains, wasting capital on transaction costs and decreasing efficiency. The efficiencies arising from e-commerce are in buying decisions, where consumers save an average of 15-20 percent through online purchase. Organizations and institutions depend upon computer based electronic technologies and networks. Measured success of performance relies on the knowledge to extend the model that motivates and communicates a comprehensive representation, combating resistance to change.

Programming is the preparation and writing of detailed instructions that negotiate operations to perform within the scope of compiled database. Computer scientists and specialists write most instructions for computers. People use programming languages that consist of words, individual letters, numerals, and symbols, as well as rules for combining those elements. A computer cannot work directly with a program written in a noncompatible programming language. Machines low-level language responds to digits which represent operating codes, memory addresses, and various symbols such as plus and minus signs. Special programs called compilers and assemblers translate programming languages into machine language. The user communicates with the smart program and the software responds and interconnects with the computer.

Intelligent shaping of a program begins with complete analysis and description of a job the computer is to perform. Programmer descriptions and interviews prepare diagrams that represent each step needed to complete the assignment. Literature, mailings, insurance, government service contracts, web page publishing, and client marketing production model the administrative focus of initial forum, important for competitive growth. Experts in this field shape information systems to perform routine tasks that appear intelligent in reasoning and learning.

Automated machines are a major reason for the computer industries' enormous triumph, although challenging issues exist within IT that speaks directly to ethical and moral conduct. In the past, as the Internet was being developed and expanded, many professionals were deeply concerned with the possibility that confidential information of a personal, financial, or even medical nature would be easily accessed by people or organizations who could take advantage of such information. Efforts to protect the security of financial transactions and shape warehoused information grow as additional populace notice the Internet. Resolutions of the technical questions of confidentiality and information protection are ethical issues that IT professionals must derive.

The question of ethics suggests that IT professionals are profoundly concerned with social issues that range from the confidentiality of personal information to bigger issues of easy access. For example, the VPN revolution benefited individuals and organizations already infatuated with technology. IT professionals demonstrate a better-than-average knowledge of skills that influence prosperity. The digital divide adds a growing significance to the industrial world. IT theorists and professionals ultimately need to resolve ethical issues beyond the related e-commerce security and control of the Internet, providing a protection to vulnerable users, free of exploitation or breaches in confidentiality.

The subspecialty of Information Architecture (IA) as one of the several possible focal points of IT incorporates system design and programming software, but also relies upon academic specialties of designing, developing, and implementing information solutions to enable competitive decision making. Invariably, this involves obtaining input from members and clients into the development of effective systems and programs that assemble real-world necessity. Shaping tomorrow's virtual society is a compilation of leading-edge computerization communications that purges information technologies and then models the impact of these technologies on individuals, organizations, and civilization.

Machine intelligence consequently refers to the technological development of platforms capable of sustaining society, although they address philosophical and ideological orientation to permeate chosen hardware and software that shape the infrastructure. The virtual society transcends local competition, extending potential boundaries of geography, and represents an evolutionary imagination, as opposed to a revolutionary change. Point and click, real-time DSL transaction integration releases the power of XML. Recognition of this transformation from a revolutionary paradigm to an evolutionary occurrence is maturation of an entire discipline and practical applications, ultimately one of the major challenges or issues confronting IT professionals today. (Appendix, Table 11)

What Can Automation Do? Focused economics of the e-commerce revolution is respect to minimizing costs of negotiating service and complying with governing transactions, then changing infrastructure design, monitoring cooperative economic policy. Strategies of profitability in doing business via the Internet minimize transaction expense with Internet Service Providers (ISP) and intermediaries. Given the dynamics of the situation, one of the key elements in cost control critical in seeking efficiency in any market is conservation of cost in routine functions.

Organizational economic theory holds that when assets are set aside for a specific objective, they simply are not used. Management wastes the assets by acting to increase the potential control benefits because goals and objectives tend to differ from the system engineers. Gaining maximum flexibility is done by allocating a collaborative information-rich environment, particularly in synchronized sequence distributions generated by IT. The new form of asset is not quantifiable in the same manner as workshop property, equipment, or other tangibles. Information technology is an economic assembly purged unlike conventional assets into a specified goal, then depleted as knowledgeable RF transmission. Because of the inherent value of such an intangible asset, e-commerce as a core business strategy is fundamentally reshaping the novelty of promotional channel sectors.

Organizations and Machines: The computerization model is operating the Internet channels and creating efficient markets through development of an expense transparency. Economists define cost transparency as a situation made possible by the abundance of free, easily obtained information on the Internet. Sellers of production, internet marketing, and collaborative distributions are more transparent to purchasers, who can then negotiate favorable acquisition prices that represent an adequate return on production expenditure.

People acquire technical skills after learning the style of experts in specific domains, gaining insight to employee resources or achievement in style of pervasive contrasting practices that influence a sense of subordinate encounter. When it becomes possible to compare the products and services of Company A than those of Company B, the net result is usually balance between expense and market fluctuations. The objective analysis of problems that cost transparency

1. Impairs salability to maintain and obtain prominent margins.
2. Limits ability to turn products and services into commodities.
3. Weaken client ethnics and loyalty.
4. Damages reputation and imply perceptions of deception.

Potential customers benefit from an environment that is extremely beneficial to alleviate resources more efficiently. The Internet potential to organizations is creating a virtually horizontal discipline for purchaser and vendor. This practice facilitates the machine efficient marketplace. Concise, efficient, and accessible knowledge anticipates the dangers of risk-free interactions made available by the anonymity of the web and addresses the ubiquitous attractions of an interchangeable acquisition of expertise. The specific mode the Internet creates or fosters as transparency expense elements include:

• How e-commerce makes a consumer quest more efficient.
• Allowance of buyer-led estimates and auctions of a bottom price floor.
• Encouragement of rational shopping by the Internet.
• Variation of prices reexamine worth structure and policy.
• Erosion of risks that vendors extract from precautious consumers.

Equally important, the e-commerce paradigm emphasizes a customer foundation over making profits. Changing the way customers think about retail cost, companies tacitly encourage dynamic incentives through price fluctuations from one marketplace to another, depending on such variables as advertisement and e-commerce conditions. Integration assembly tools remove the risk factors of integration projects.

Creating Value in the Internet Era

IT professionals develop systems rapidly and are still in a pioneering stage of emerging frontiers in the VPN technology. Included are integrated circuits of revelations, object technologies, knowledge, distributed intelligence, and evolving multiagent intelligent systems. Satellite networks and mobile commuting are the virtual cooperation of the new millennium. IT is improving the production of hardware semiconductor technologies and expanding novelty into automated systems.

The future theories of hardware and software systems include networks, capable of figurative distributing intelligence, reviewed by a National Science Foundation (NSF). IT professionals represent the next generation of telecommunications, intelligent databases, collaborative technologies, and high performance computing platforms. Formulating vigorous ad hoc decisions about scarce resources challenge financial and human deficiencies allocated to these frontiers, creating value to organizations and policy maker issues. The principle is that time and space are no longer the organizing foundations that they once were. A virtual organization does not exist in one place or perhaps even one time, but whenever and wherever the merchandise happens to be. Micromanagement extends the value of tools to a full range of new products.

Creating worth in the e-commerce era defines membership as a somewhat exclusive club. Information sciences distinguish no boundaries and are universal to be ethical. At the heart of VPN, technologies are valid concerns of understanding problem solutions and representing knowledge structures that support the construction of valuable meaning. Differences relate to the outflow of serving new consumers and catalogues of variations in the ways that clients value the organizations exchange.

Everything ultimately revolves around serving a consumer well. Internet users mandate the recognition that arrives in all shapes, sizes, and varieties. IT professional institutions or channel organizations become imaginative within the fabric of the nation's educational paradigms, as the era of Internet users define technology commonplace.

The ability to remember the routine risks of reality preserves the understanding of background awareness, which enables a realistic perception of a more refined sensitivity to the mode-shared situation. A tendency to respond positively sustains our interpersonal efforts that depend on the ability to complete the task. Serving the needs of a socioeconomic condition or status within the IT revolution portray a critical moral ethical apprehension by means of economic implications as well. Each of these variations can foster the emergence of efficient markets in resources and customer desires equitably.

Summary of Chapter

The analysis identified a number of the critical issues that directly reflect upon IT professionals. The literature reviewed the researcher's implementation discipline that expands exponentially each year. IT specialists learn to model a sensible direction to a perceived goal that can emerge new questions. From the ethical perspective, a code of secure research and development consists of tacit needs of policy integration. IT is at the crossroads of material technologies, shaping the infrastructure of daily routine actions that eventually model the paradigms of discovery for the next several years.

The component theory discussed will certainly increase the IT perspectives of security professionals who are expanding the internet marketing opportunities or practice logical self-monitoring protocol of assigned responsibilities to each system user through channeled administrative supervisory control. Integrating human intelligence is the fundamental science of creating standards of information distribution to secure integrated data systems, while improving on the collaborative practices for educating IT professionals, rationalizing the economic effect or ensuring ethical access, purging developmental directions and policies. Knowledge is a vital tool to improving enterprise customer service.

Chapter 5

Summary, Findings, Recommendations

Summary

This chapter provides a review of the analysis results and supplementary recommendations and conclusions drawn from the application design. Specific findings of the firewall protection considerations including office automation obstacles provide hypotheses support. Implementation results expose recommendations that may prove useful for future projects of resource management or VPN implementations.

The goal of the review is to provide general guidelines for application software development and security concerns that establish membership protection, developing feasible tasks in a real-world application. Data acquired during the study demonstrate information protections which could become security vulnerabilities that integrate into the UNIX user authentication to the registrar server, which provides a transparent client/server interface routine business operations. Career opportunities for IT professionals are among the most technically specialized experts of computerization who design the circuits engraved on chips and the wiring that programmers use to correct alteration.

Specific Findings

IT is enhancing corporate information channels to extend beyond traditional client services, adding novelty to a marketable product of knowledgeable distribution segments. Deployment of high-level VPN technology that is effective retains a volume of market research reports or project-based evaluations valuable to the consulting industry of professionals and future contingencies. The changing legal threshold of technology deployment in the national workplace present intriguing issues to complicate privacy and intellectual copyright perspectives, as teaching employees emulation technology becomes the norm.

Interactive machines mimic human intelligence to improve capabilities in pattern matching and to diagnose methods to describe object events. AI allows computers to operate mechanical devices efficiently, achieving perceived business expectations that improve member support of daily business activities—a capability that is not usually accessible to traditional commerce operations. Today's prototype of a virtual society is an interfaced environment that incorporates a large knowledge-based spectrum of a specific problem domain. The analysis identified the application of IT to the business arena, specifically to the shaping and value of e-commerce knowledge.

Review of Findings:

- Communication facilitates information and user feedback.
- Goal = to expedite the flow of data accuracy in a timely manor.
- Distribute a VPN to shareholders to meet model decisions.
- Develop the electronic information warehouse.
- Update developmental resources and Web site.
- Shareholders restructure elements and micromanage activities.

Direct Client Billing Proposal:

- Goal = to reduce processing time and introduce less billing errors; expedite payments; and ensure fiscal accountability of service quality through micromanagement.
- Determine system enhancements and feasibility of future changes necessary to address the developmental requirements.
- Modify recommended changes integrated to services program to become a repository of information for reports and user monitoring.
- Serve as a demographic database for members though quality control development.

Rationale for Attacking Consumer Systems:

- Use the client-side launching points for further attacks
- Trade illegal software

- Excitement in testing skills
- Distributed denial-of-service attacks
- Gather personal information
- Destroy a user's system
- Revenge

Windows and UNIX Server Capability:

- E-mail facilities on the Internet, DSL, T1 connections
- Presentation facilities for delivery of multimedia support
- Desktop publishing
- Retainer analyses and market research
- Firewall gateway surveillance

Office Automation Obstacles

In tallying, enterprise companies operate their own internal networks. Conversely, the issue is not merely whether computerized activities are possible but if information becomes so numerous or a vital linkage indispensable to everyone. Imagine, for example, trying to construct a single new service that would tap into the corporate databases of all the local competitors and analyze the lowest price for a desired system. Computer databases maintained are usually out of consumers' reach. In addition, even if access is available for the generic data, the important question to answer is what IT professionals can do to remedy this chaotic state of affairs.

Information can take the form of text, drawings, music, speech, photographs, stock, invoices, software, streaming live video, and numerous other phenomenons. Once information is computerized, the data becomes a deceptively uniform sequence of zeros and ones designed flexible in transport streaming, able to carry varying degrees of speed, accuracy, and security to match dissimilar computer capabilities and essentials. Transporting flexibly is not enough since computers share no universal knowledge concepts. We therefore set widely understood common communication conventions.

Finally, the truly useful infrastructure should be equipped with common servers that provide a few basic information services of widely computerized interests. Surrounding the rudimentary communications principals design the millions of mainframes programmed to buy and sell information services. Superficially, the telephone system, with millions of glass fibers wires, is reaching every home and office and would be adequate for shaping the database scheme. While this may become possible in the future, that vision is not a reality today.

The TCP/IP limitations stem from a century-long dedication to a specific task-transmitting voice to human listeners, as individual speech sounds vibrate at frequencies of no more than 4,000 cycles per second, with matching sensitivity. However, a computer carrier signal built on a wider range of vibrations is ranging from a few hundred characters per second for PC text, to the equivalent of a few million characters per second for a supercomputer forecasting familiar graphical patterns.

Tomorrow's multiprocessor machines promise to increase the envelope of this range even further. TCP/IP composed largely of copper wire has mixed speed capabilities and is essentially incapable of high-speed video images transmission, although it can broadcast an array of long-distance multiplex glass fibers; integration can significantly reduce product acquisition and project start-up costs.

The assortment of transmission bandwidths to meet extensive requirements motivates consumers to pay only for essential bandwidth. The importance of reliability in transporting information from a server computer continues to shape and anticipate the total production output potential that could effectively quantify a small portion of the object representations. Given such a small margin for error is devastating in transmitting forecasts of financial data, software, or schematic diagrams. To resolve this issue, the telephone companies offer special, cleaner lines, in addition to knowledgeable recourses to ensure totally noise-free transmission, automatic resend, and then check of data.

Senders of legal contracts, sales proposals, and new product designs scramble their message to ensure privacy. However, the levels of bandwidth, reliability, scalability, and security require reasonable charges determined by the meaning of a flexible transport, providing an open-end versatility. For example, the phone companies offer transmission for a "T1" service, which carries the equivalent of twenty-four simultaneous voice conversations, into a much faster "DS3" service, which transports the equivalent of twenty-eight T1s. There is no reason why tomorrow's telephone system could not offer these adjustable services. Computers and other gear for routing messages would have to accommodate the wide range of data rates and other capabilities.

The Integrated Services Digital Network (ISDN), in practice, exists in a narrowband form. Narrowband (NISDN) suffers from the same constraints as classical voice telephony in many ways. For example, NISDN can carry information only in fixed chunks of 64 kilobits per second and accommodates the greater demands of the industry. A newer approach called Broadband ISDN or (BISDN) seems promising as an entirely different architecture. From interpreting, the Consultative Committee on International Telephony and Telegraphy (CCITT) appears to have reached a consensus for the need of BISDN. The idea is conceivable that BISDN, employing 150 million-bit per second chunks would come into use during the next decade.

The online penetration of legal issues, privacy, and infrastructure ownership extends the novel characteristics of boundaries beyond all the demographic levels of race, geography, income, and education. At the same time, there is evidence of a growing digital segregate, with respect to both computer data ownership and Internet access. Ownership empowered access is growing among the prosperous and well educated; essentially, individuals of diverse education are part of a stable household and among the largest demographic e-commerce sectors. Income brackets usually of more than $75,000 a year enjoy DSL. Of households at the lowest income, under $10,000, only 8 percent acquire a computer each year.

Recommendations for Further Research

The importance of adding novelty to marketing strategy often is cultivating a visual presentation position, which includes a camera apparatus and a well-defined environment of relationship recognition, although each evaluation will interpret scenarios differently. The fact is automated VPN technology cannot be approached lightly because it is not for sale. The discussed issues of implementing IT in the workplace will certainly summarize the basic business model of tested and proven methodologies, improving telecommunications operations and cultivating important financial objectives.

Each topic of this evaluation introduced the most import issues of the researcher's real-world experience in today's managed information systems. A precise review of in-house efforts describes the eventual development of e-commerce operations, complying with legal obligations to scrutinize associates and highlight implementation benchmarks.

The vast number of VPN channel highways transport RF carrier waves, in conversations on any topic, and support electricity appliances. The financial, human resources, and time management activities are important to consider before beginning a possible execution.

Contrasting these observations, workers frequently complain they can no longer perform their job activities effectively for fear of legal or discipline consequences. According to the Equal Employment Occupation Commission (EEOC), employed personnel should not feel threatened in the workplace and must feel comfortable in speaking about humiliation or harassment on the job. Determining the correct amount of contingent coverage to handle twenty-four-hour operations requires careful analysis of the emerging corporate perspectives. Executive leaders balance the necessary security captive and determine the required outsourced representative responsible for each operation. In addition, added security management is monitoring streams of multichannel binary environments, gaining a better perspective to what extent libelous acts prove negligent. IT professionals must amend from region-to-region as the boundaries are expanding novelty and changing government regulations.

Computerization is a truly useful part of society when designers model the infrastructure into a virtual advertisement of information services, with occasional forays by the daring few along uncharted, unpredictable, and treacherous roads, full of unspeakable security apparatus. The modern enterprise advocates issues of infrastructure ownership that reinstates IT professionals to unite through the Internet.

Executives understand procedure as accounts focus is bottom line and attorneys continually remix the hypothetical scenario of misfortune, as the emergence of information e-commerce becomes apparent. Perhaps the value is the mark of novelty in the exchange of information among widespread users of itinerant isolation.

The foresight of this review is for the computer experts in every region to interconnect moving text, pictures, audio, movies, software, and designs of useful means rapidly, making possible an unlimited number of new activities, which should improve our economy and mode of vivacity. When confronting communications issues with enterprise workers ineffectively, we reason through arcane and arbitrary procedures.

Soon the National Information Infrastructure (NII) will become a common resource of computer-communication services, as easy to use and important as the TCP/IP network, the electric power grid, and the interstate highways. Economic potentials certainly measure and extend the future VPN possibilities. The information system delivery tends to calculate financial impact of import business decisions that technology research does not cover in every obstacle or deployment project operation, although presents future quantities of objective, "Target Market." To step back from aspirations, with their enormous potential of uses, present the objective benefit of resolving vision into reality.

For example, the routine delivery of mail in five seconds, instead of five days; where advertising could be done in reverse, with consumers broadcasting needs to suppliers; where goods would be ordered and paid for electronically; employees at outlander offices taking care of children at home; or enriching assortment of interactive. Designers and marketers collaborate on product planning, even though they are a continent apart and not able to meet at the same time. Many of these ideas have become realities, although their practicality is not for everyone.

The extending router products are a unique range of structured tools. A common misconception arising from the wealth of computers and communications is that we already have an information infrastructure in position. Nevertheless, to qualify as infrastructure, a resource must be widely available, easy to use, and inexpensive. Moreover, it must serve as a foundation for potentially unlimited uses. Roads, telephones, and power lines are everywhere and accessible by almost everyone for a modest expense without effort. Unfortunately, the convenience of sequenced wall sockets throughout the nation's offices and homes, into which computers can exchange a multitude of diverse information, does not exist. This characterization may seem unduly to members who connect to the phone line to exchange electronic mall, shop, bank, or tap into the wealth of financial or social stratum.

Several major revolutions in mathematics and technology over the last five hundred years inspire revision of the enterprise information system altogether. Inventive focus is a fundamental misunderstanding of computerization trends. The inventions that modify and enhance the techniques of proof, clearly alter the underlying philosophy they prove or distinguish from a logically coherent story. Deriving scrutiny is the skeptical expert real-world descriptions that stimulate new mathematics, survives and connects imperatives of the outside world. Revolving the enigma of human imagination is a two-way traffic of ideas that helps resolve significant problems for the future.

"Moore's Law" accurately illustrates that computer development doubles every eighteen months. To expand on his "empirical law" is to continue without an agreed theoretical underpinning of national ethics and underscores software development that grows faster than Moore's Law does for over fifty years. As technology, progress continues along the economic benchmarks, which slow growth tendencies and expect exponential progress. Because of these facts, in the equipment world, software, not hardware, determines the state of the art and dictates the future pace of change. Machinery moves forward on invented circuits that significantly rely on software architectures to support and build novel hardware relationships.

The state of the network environment records efforts of creativity and collected information to handle and find positive solutions to problems. To secure the remaining issues with confidence and inspiration is to create a superior situation. Priorities entail helping the development of structural changes and committing them to the path of the rule-of-law for filling the pledge of the corporate challenge. Government organizations and the media argue a variety of good sense solutions of selectivity to address and confirm concerns that may paralyze our reasoning. Through imperative judgments, we regain the ability to prioritize the numerous worthy causes.

Effective software integration is dramatically reducing the need for custom integration code. The IT professionals discipline is often prepared by establishing tendencies that supplement each other. Psychologically we have inclination to underestimate large risks and overestimate small ones. The multimedia tendency is to focus on the dramatic rather than resolve everyday risks. Attracting public interest into the present, we find tragedy and accidents that reflect in mortality statistics. Through phenomenon, a variety of reasons becomes an astounding intervention built upon the premise of how much wealth accumulates. Merging together systems and structural members continue to inspire upgrades and replace obsolete technologies that integrate significant nobility.

Implementing visual technologies extend senses into a useful and enjoyable novelty that augment and fully express potential capabilities. The human consciousness of today's information systems rely on a video camera apparatus with a field of view of about 160 degrees horizontally. In contrast, the human retina represents one hundred million brightness receptors and about five million color cone receptors. The complexity and morality of the challenges impose an unrecognizable hindrance from the twentieth century system architecture. The historical boundaries of the last five hundred years is built upon prejudices, with willingness towards blind alliance, as scientific revelations help organizations to understand and appreciate technological innovations of what truly is a potential for unique change.

The Internet explosion represents ingenious failures of the AI science because of the gradual economic setbacks in the human resource potential. The media target demographics further implicate a disturbing tendency of motivating computer end users. Computerization may exhibit augmenting capabilities such as megabyte memory application or provide accelerated calculations that embrace security intelligence to move faster than polarization, adapting a scheme beyond what individuals can easily ascertain, or obtain full data transmissions within intuitive graphical browser that determines a desired collaboration between machines and the populace.

To take advantage of the emerging web presence, AI must learn to reason, as the business community learns to behave more like machines. The Internet brings the very essence of expertise that generally produced devices that serve already recognized behavior, or defines undiscovered effects based on commerce friendships, supported by physical proximity. The same day-to-day business environment is less likely to value change, as the technical understanding strives for convincing conversation, making casual discussion more difficult, and rendering future support less applicable, likely to diminish strength ties of physical proximity.

The growth of the Internet has brought increased attention to individual privacy as the antiquity concepts encoded a chip, a caller, or a semi-sentient machine to accomplish restricted access to personal electronic backgrounds. Information access is probably the most significant benefit of the Internet revolution and related discipline technologies. The telephone modes of communication offer advances, although mandates adjustments to the Internet-ready lifestyle. The pace of present day technologies necessitates a resourceful controlling factor and audit tracking sequence. Emerging technologies extend access to a ruling personal infrastructure, as hackers target individual systems looking for data resources of potential client control. Network auditing offers a unique coverage of the entire spectrum of network problems and issues. Annex monitoring will certainly detect an unauthorized intrusion and collect evidence on the IDSN audit log. In addition, the firewall can monitor activity by examining each packet of the data. Filtering and monitoring benefits of management operating systems protect privacy that is sensitive to intruder attack, triggering defenses by means of scanning and encryption technology, although it cannot stop every attack. A virus microfunnel approach can examine e-mail, open files and attachments, or initiate a scan of downloadable files. Sensible precautions enable a comfort level of encryption protection for confidential information, effective for knowledge-based reason, vital in a binary transmission.

Computer researchers are learning to integrate the laws of life with electronic circuitry, creating new environments similar to nature. The underlying *principles* revealed from knowledge evolution are adapting computers to operate evolutionary algorithms to help shape the future of software programs. Using processes closely related to the natural sciences digital neural networks will evolve new spins to callers as the "intelligent" decision-making behavior mirrors a sensible progress. Computer scientists define principles that extend to the next generation of computer software and hardware architectures, as digital agents help locate resources and destroy computer viruses. The latest developments of information distribution intrigue a glimpse into the future of technology.

Bibliography

Amoroso, E. (1994). *Fundamentals of Computer Security Technology.* Upper Saddle River, Prentice Hall.

Awad, E. M. (1996). *Building Expert Systems*, West Publishing Company.

Bleeke, J., Ernst, David, Ed. (1994). *Collaborating to Compete: Using Strategic Alliances and Acquisitions in the Global Marketplace*, John Wiley & Sons.

Cascio, W. F., Zammuto, R. F. (1987). *Social Trends and Staffing Policies*, Denver, University of Colorado Press.

Collins, H., Kusch, Martin (1998). *The Shape of Actions: What Humans and Machines Can Do*. Cambridge, The MIT Press.

Connolly, T., Begg, Carolyn (2000). *Database Solutions*. California, Addison-Wesley.

DeVore, P. W. (1992). Introduction to Transportation Technology. *Transportation in Technology Education*. J. R. W. S. K. (eds.). Columbus, Glencoe: 1-32.

Erwin, M., Scott, Charlie, Wolf, Paul (1999). *Virtual Private Networks*, O'Reilly & Associates, Inc.

Giarraputo, J. (2001). Service Providers Lock Down Internet Security. *Global Finance*. Westchester. PA: 77-84.

Harrington, J. L. (2000). *Object Oriented Database Design*. San Francisco, Morgan Kaufmann Publishers.

Bibliography Continued

Held, G. (2000). *Voice & Data Internetworking*. New York, McGraw-Hill.

Jones, V. C. (2001). *High Availability Networking with Cisco*. Boston, Addison-Wesley.

Keen, P., McDonald, Mark (2000). *The E-Process Edge: Creating Customer Value and Business Wealth in the Internet Era*. Berkeley, Osborne, McGraw Hill.

Noe, R. A., Ford, J.K, Ed. (1992). *Emerging Issues and New Directions for Training Research*. Research In Personal and Human Resource Management. Greenwich, JAI Press.

Parnell, T. (1999). *Building High-Speed Networks*. Berkeley, Osborne & McGraw-Hill.

Pyke, M. (1957). *Automation: Its Purpose and Future*, New York Press.

Rao, R. S. (2001). Counterspy. *Forbes*: 1-2.

Rolfe, J. (2000). Peering Into the Broadband WAN Future. *Communication Systems Design*: 23-30.

Sheaffer, D. (2000). Delivering High Availability. *Embedded Developer's Journal*: 50.

Simmons, K., Buse, Jarret W., Halpin, Todd B. (2000). *Network Design*. Scottsdale, Coriolis.

Skyrme, D. J. (1999). *Knowledge Networking, Creating the Collaborative Enterprise*. Woburn, Butterworth-Heinemann.

Spengler, O. (1922). *The Decline of the West*, New York Press.

Sprague, R. H. J., McNurlin, Barbara C. (1998). *Information Systems Management In Practice*, Prentice Hall.

Bibliography Continued

Truxal, J. G. (2001). *Automation and Jobs*, State University of New York Press.

Wilson, J. (1989). *New Technologies and Social Invention*. Jordanstown, University of Ulster.

Zuckerman, M. B. (1995). Where Have the Good Jobs Gone? *US News and World Report*. 119: 60.

Professional Affiliations and Achievement Awards

Studio K Productions LLC, SonicImpact, IEEE Computer Society, Association Computing Machinery (ACM), Computer Security Institute (CSI), ABI Research Fellow, Network World, Decision Analyst, Technology Advisory Board, Trump University, Data Center University, Network Solutions Gold VIP, Microsoft Partners, Siteserver, 1and1 Webhosting, Techsay Research Panel Pinnacle, Oracle, Cisco Systems, Sun Developers, Berklee Music, South Florida Federation of Musicians. Additional Literary Work; Electronic Design: The Myth of the Shrinking Design Cycle. The State of Analog and Digital Recording: Perspective's of a Broadcast Engineer. Awards; Who's Who in America, The World, Entertainment, Finance, Contributions in Recording Arts, NRI High Honors, International Society Photographers Hall of Fame and Editors Choice Awards.

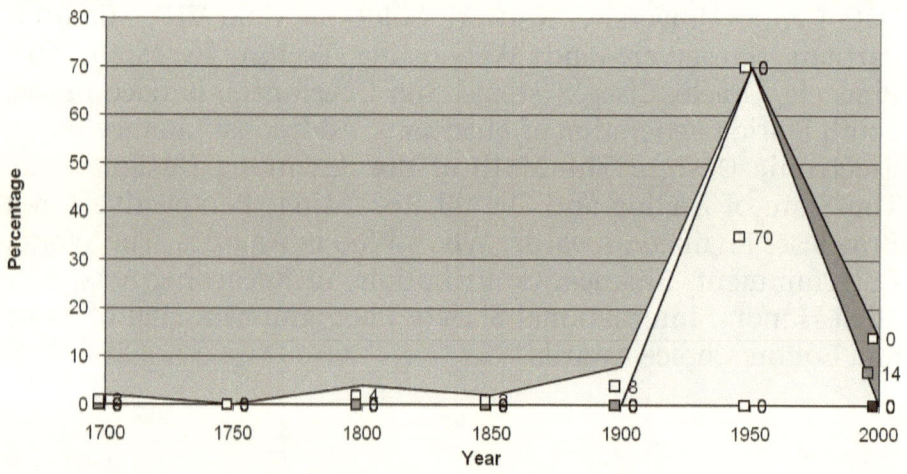

Three-Hundred Years of
U.S. Technology Advancements

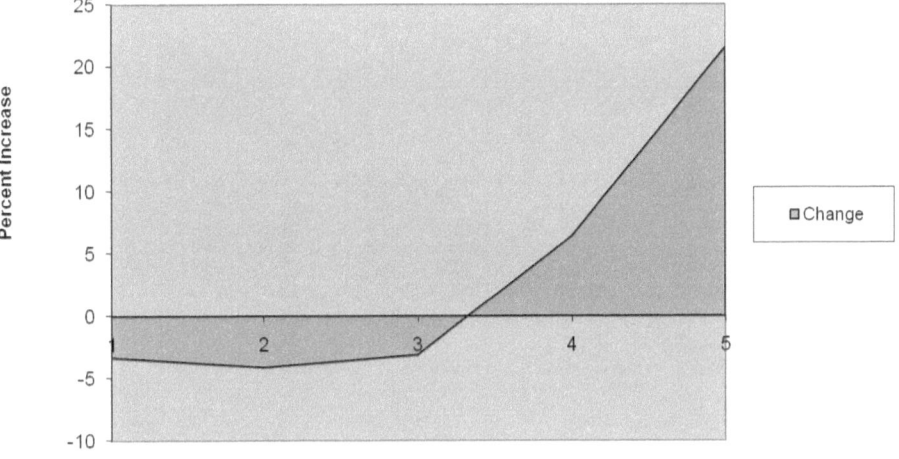

Tax Analysis from October 2000-2002

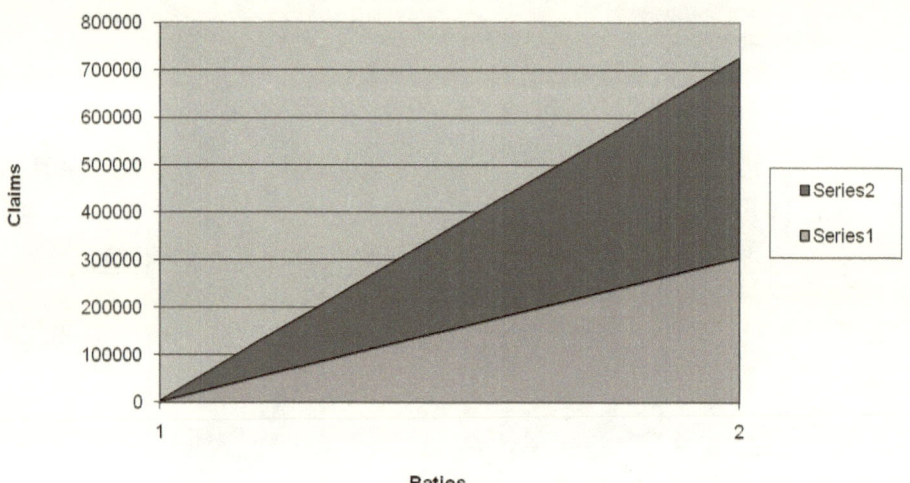

Figure 2B First Quarter Unemployment 2000-2001

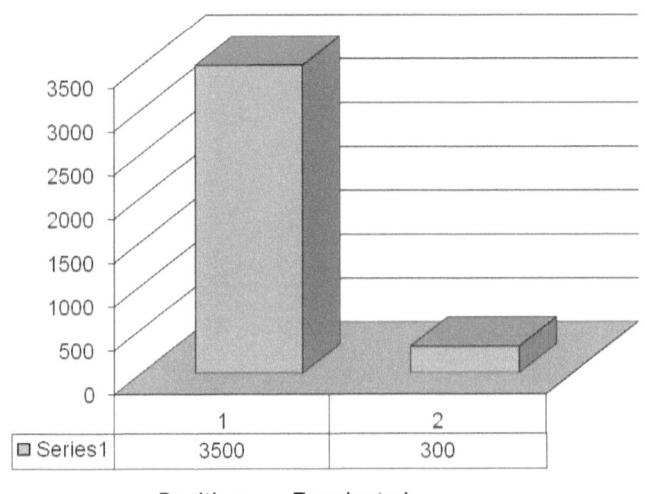

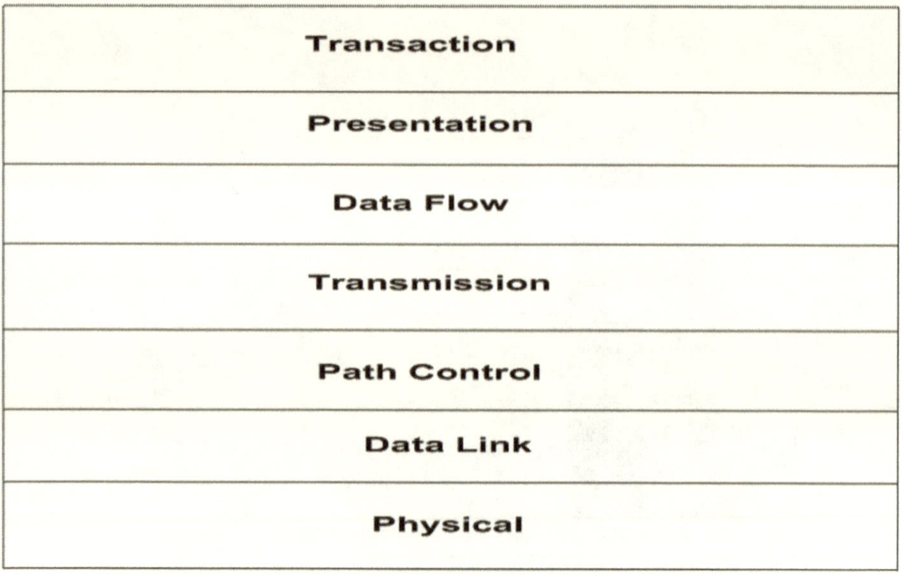

Transaction
Presentation
Data Flow
Transmission
Path Control
Data Link
Physical

The SNA is IBM's proprietary terminal-to-mainframe protocol,
introduced in 1974. Describes a seven-layer system, computer-to-
with each layer building on the services provided by the previous
layer. Devices on system are connected on a Synchronous Data
Link Control (SDLC) protocol, running on serial lines. SNA is not
compatible with the OSI Reference model.

Figure 3 Common Systems Network Architecture

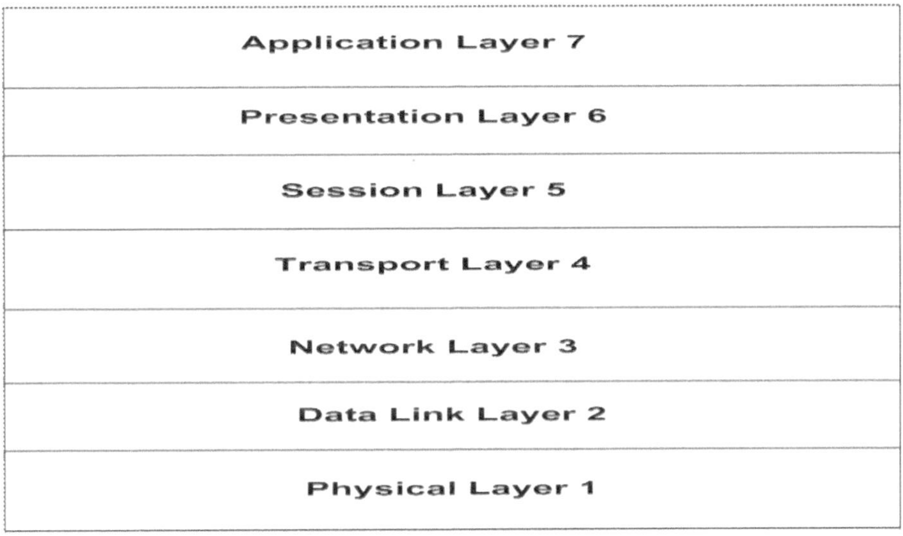

Application Layer 7
Presentation Layer 6
Session Layer 5
Transport Layer 4
Network Layer 3
Data Link Layer 2
Physical Layer 1

The OSI builds on the seven-layer system, computer-to-computer communications, with each layer building on the services provided by the previous layer. Devices on system are connected on a Asynchronous Transfer Mode (ATM). OSI is not compatible with the SNA Reference model.

Figure 4 International Network Architecture

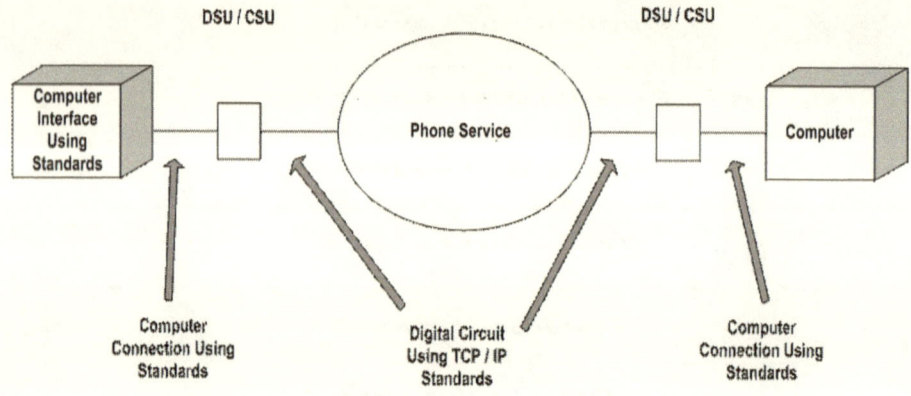

A digital circuit with a DSU/CSU, converts between the digital standards in a telephone system and those used by the computer lindustry. Digital circuits are leased from common carriers and are the fundamental scheme for long-distance networks. Each circuit extends beyond two points. The transmission cost depends on the circuit capacity and distance.

Figure 5 Data Services Unit / Channel Service Unit

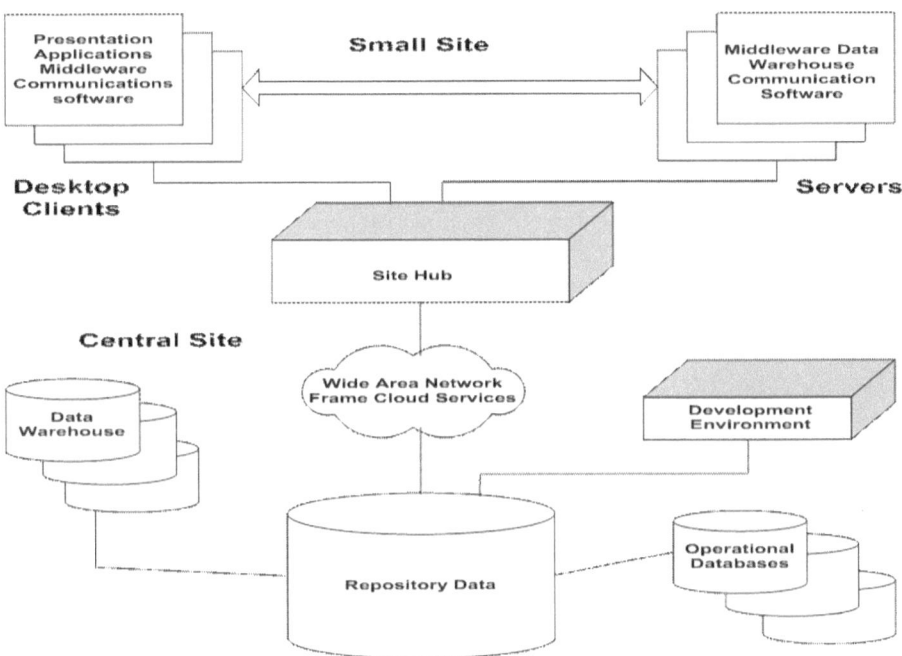

Figure 6 Management Input-Output Process

© 2002 by Carl Catalano

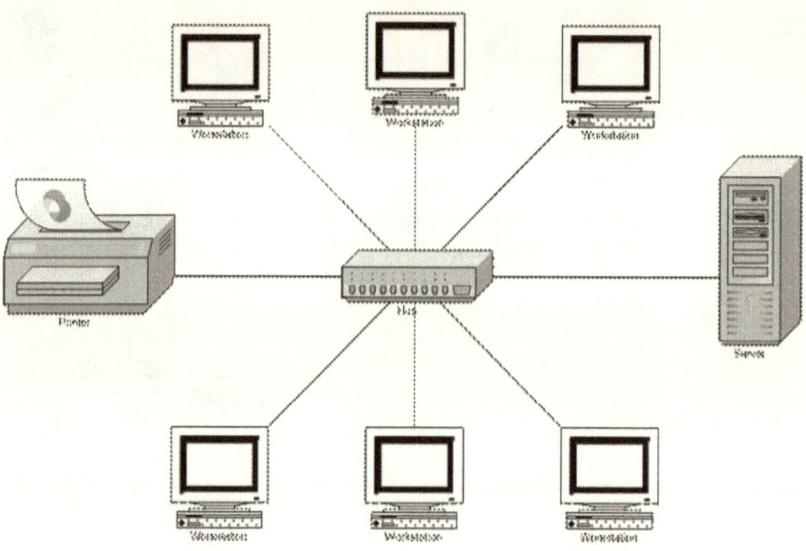

Figure 7 Average Star Computer Network

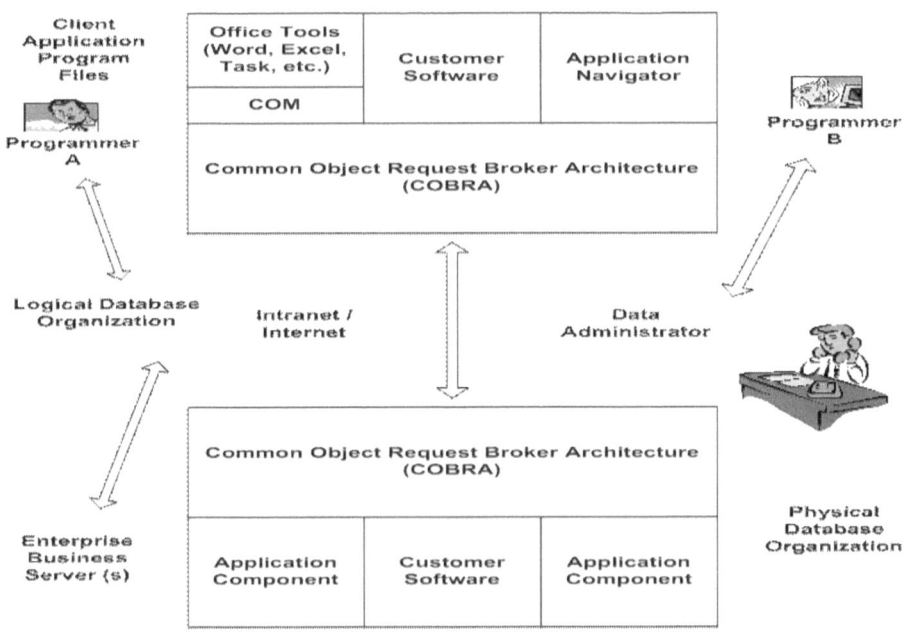

Figure 8 Application Software's New Architecture

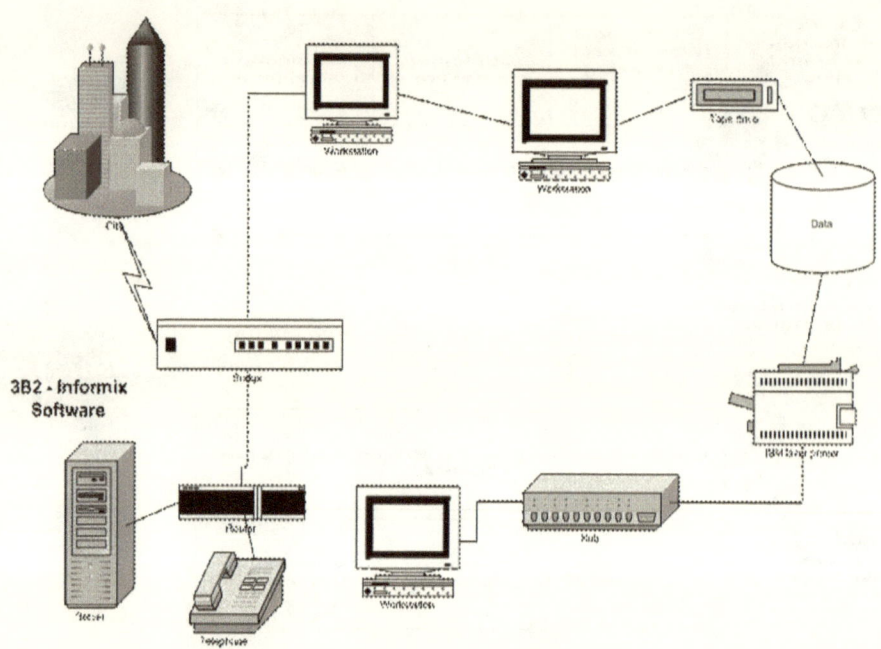

Figure 9 Original Framework LAN Based System

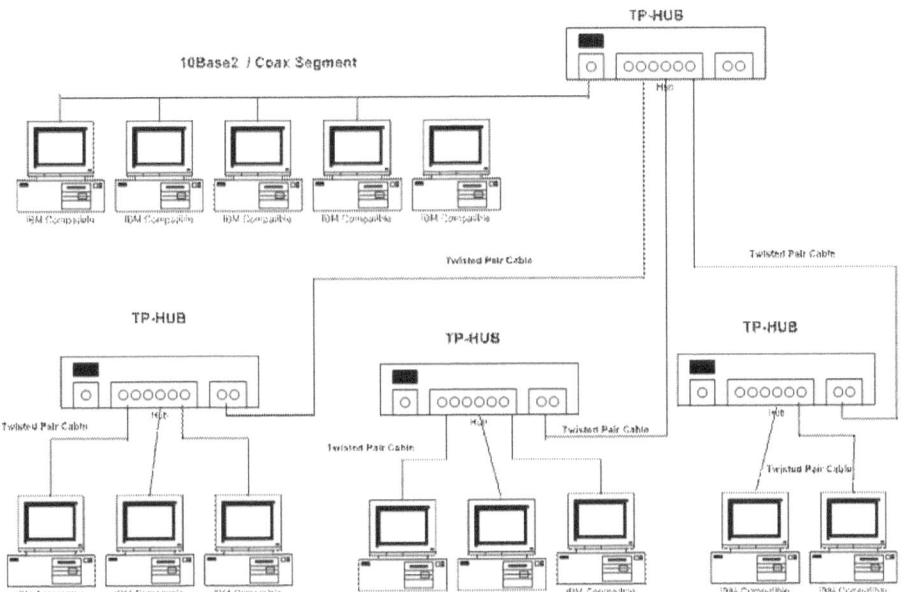

The cable-extension method of connections has a maximum throughput limitation. Hubs and repeaters are non-intelligent devices. Port is amplified and sent out to ALL ports. network slowdown due to cable-segments on the network. One network transmission fills up-flow into ALL cable-segments in ONE active state.

Figure 10 Original Cable-Extension Method

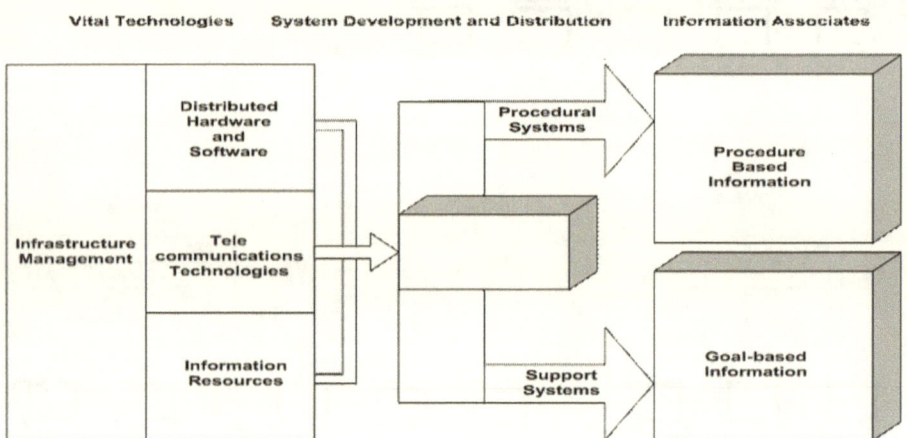

Figure 11 The New DNA Distributed Scheme

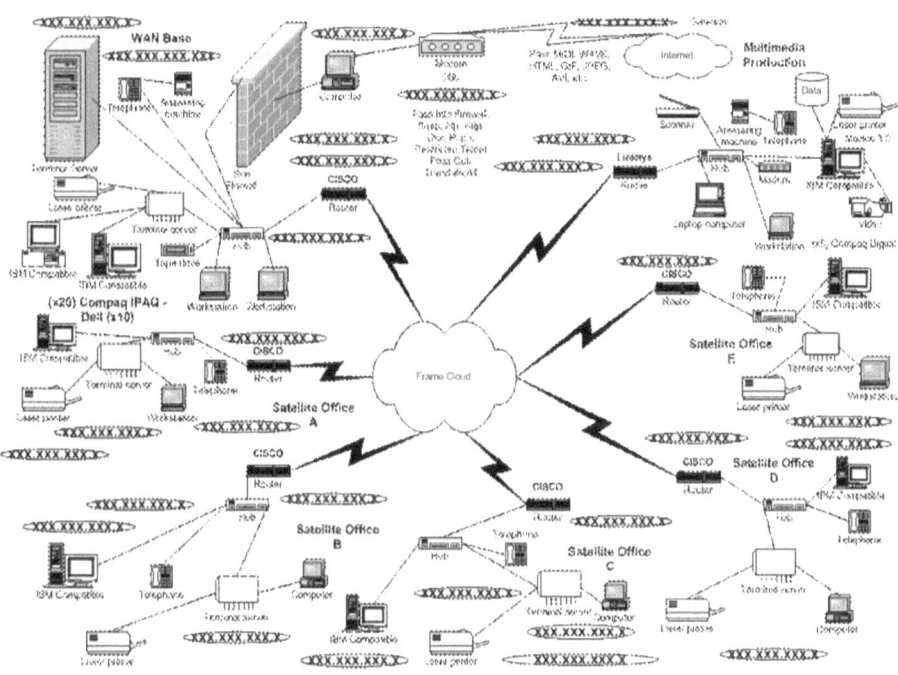

Figure 12 Implemented VPN Architecture

© 2002 by Carl Cataldone

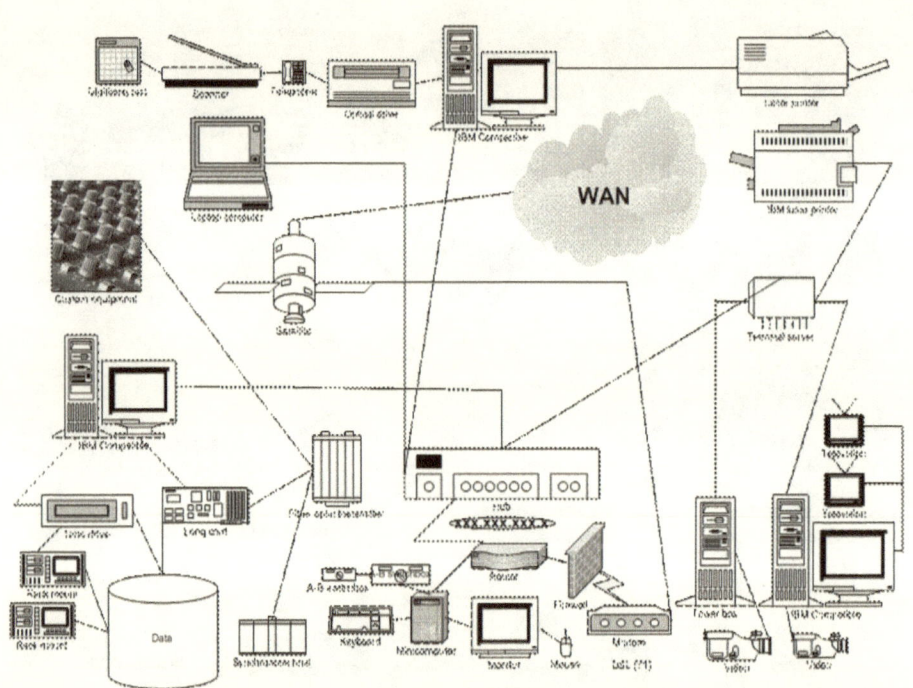

Figure 13 Implemented Multimedia Production Process

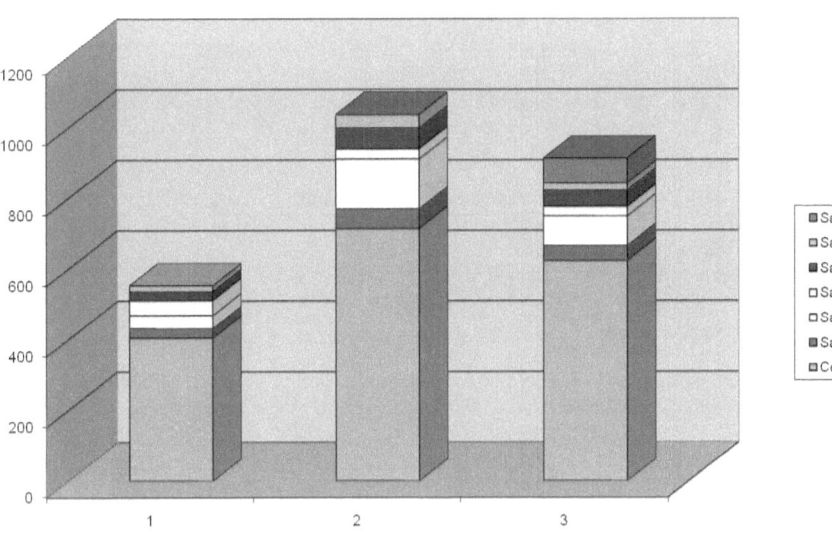

Figure 14 Three-Year Growth Analysis

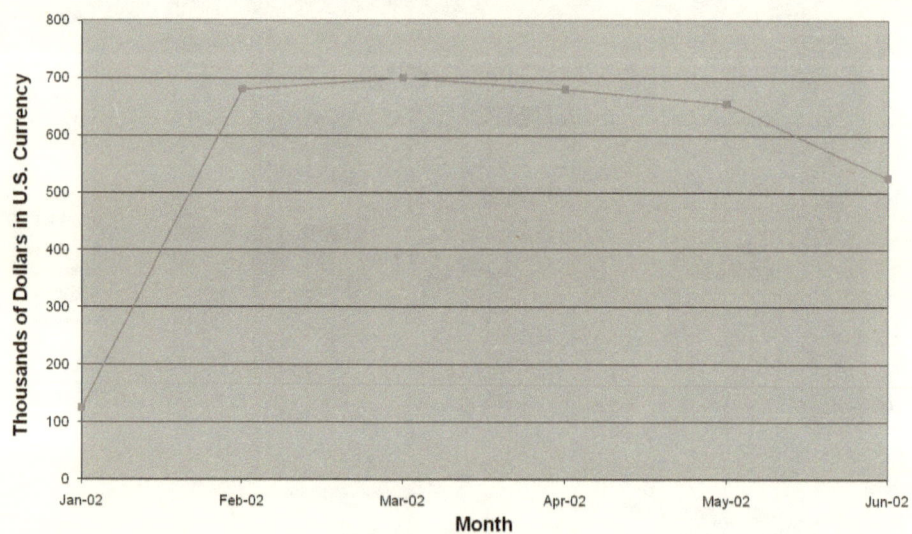

2002 Resource Review

	1700	1750	1800	1850	1900	1950	2000
100							
90							
80							
70						70	
60							
50							
40							
30							
20							
10							14
	2		4	2	8		

2000	304000	March		Tech	Stock	-74.4	Nasdaq
2001	421000	March		April	2001	-80.03	DowJones
Increase	117000						
Claims	404500	April			3500	300	Motorola
Highest	165564	April					
Telecom	26464	April					

	Tel-Res	Tel-Bus	Wireless	Cable	Satellite	
Dade	10.5	18	17	11.5	6	
Broward	9.5	16.5	16.5	11	6	Current
Collier	2.5	9.5	9.5	11	6	
	Tel-Res	Tel-Bus	Wireless	Cable	Satellite	
Dade	7.19	14.29	14.29	14.29	13.17	
Broward	7.29	14.09	14.09	14.09	13.17	New
Collier	4.71	11.51	11.51	11.51	13.17	
Dade	-3.31	-3.71	-2.71	2.79	7.17	
Broward	-2.21	-2.41	-2.41	3.09	7.17	Difference
Collier	2.21	2.01	2.01	0.51	7.17	3 County
Change	-3.31	-4.11	-3.11	6.39	21.51	
Average	8.3808					

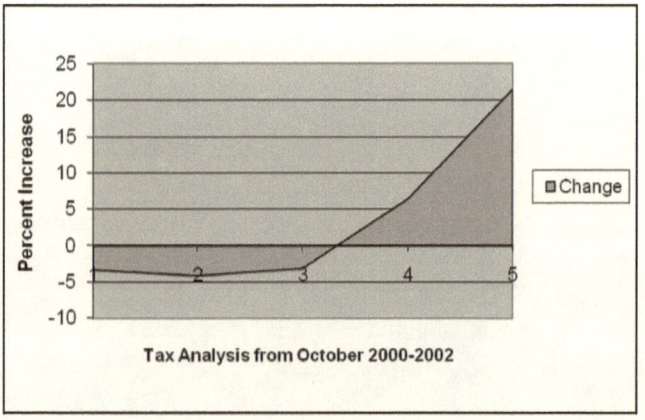

Tax Analysis from October 2000-2002

Menu   Staff Data

F1	on call		
F2	search		
F3	off call		
F4	update		
F5	search	discipline	A
F6	search	discipline	B
F7	search	discipline	C
F8	switch	view	
F9	staff	history	

Client Data

F1	confirm	order	record
F2	search	not	confirm
F3	query	by	code
F4	update	code	field
F5	display	Current	details
F6	creates	relief	record
F7	list	available	staff
F8	next	available	window
F9	assign	staff	name
F10	display	client	history
F11	display	client	matrix

Available Staff

F1	left word	comments date	
F2	no ans.	comments date	
F3	off call	comments date	
F4	Will Call	comments date	
F5	queries	staff	available
F6	queries	discipline	available
F7	updates	staff	available
F8	switches	next	window

MIDI

Name	IP Address	Brand	Port	Model	Channel	Descript.
CorpRec	192.168.100.47	Olympus		DS-2300		PhoneRec
IT 01	192.168.100.48	MAC		Powerbook		Notebook
IT 02	192.168.100.49	HP		DV9000		Notebook
Automate	192.168.100.77	MAC x4		G5		Audio-Vis
Control	192.168.100.96	Korg		EXBRadias		AnaSynth
RacServer	192.168.100.97	Systemax		V133		TB Storage
C2620	192.168.100.98	Linksys		SRW2024		Router
FW	192.168.100.99	Webview		V1.2		Firewall
Server	192.168.100.100	Systemax T1	DSL	Enterprise		Services
Mix	192.168.100.101	Mackie	7	DBX200	1	AutoMixer
Sequence		Roland	7	MV8800	1	AutoSubm
HDR		Digidesign	8	ProTools	1	UnlimTrk
PC 37	192.168.100.102	HPTouch		TL52 64x2		Workst. A
PC 38	192.168.100.103	HP		4650		Workst. B
PC 39	192.168.100.104	Sony		CLS30		Workst. C
PC 40	192.168.100.105	HP		xw9300		Workst. D
PC 41	192.168.100.106	HP		xw4400		Workst. E
UnitorATM 8		Emagic		U8-ATM8		MIDI Inter
MIDI		Digidesign		8x8		MIDI Rout
Link		Linksys		WRVS4400		VPN
CDR		HHB		800		CD Recr.
Wave Rec		MOTU		HD192		HDR-Sync
MC		Roland	3	MC-808	15	GrooveBx
MV		Roland	5	8800	14	WaveP
ONYX		Mackie	3	400	3	WaveM A
Ultra		Behrhinger	3	UcurvPro	12	WaveM B
Sample		Roland	5	SP-606	8	WaveM C
Synth A		Yamaha	6	TyroP800	5	Keybrd A
Synth B		Roland	6	G-70	6	Keybrd B
Synth C		Korg	1	Oasys	7	Keybrd C
Synth D		Alesis	1	Fusion6	9	Keybrd D
Synth E		Korg	1	Radias	10	Keybrd E
Synth F		Korg	2	M3	11	Keybrd F
Synth G		Yamaha	4	Motiff XS6	13	Keybrd G
VocalPro		DigiTech	7	Vx400	2	VEP

Name	IP Address	Brand	MIDI Port	Model	Channel	Descript.
CorpRec	192.168.100.47	Olympus		DS-2300		PhoneRec
IT 01	192.168.100.48	MAC		Powerbook		Notebook
IT 02	192.168.100.49	HP		DV9000		Notebook
Automate	192.168.100.77	MAC		G5		Audio-Vis
Control	192.168.100.96	Korg		Radias		AnaSynth
RacServer	192.168.100.97	Systemax		V133		TB Storage
C2620	192.168.100.98	Linksys		SRW2024		Router
FW	192.168.100.99	Webview		V1.2		Firewall
Server	192.168.100.100	Systemax T1	DSL	Enterprise		Services
Mix	192.168.100.101	Mackie	7	DBX200	1	AutoMixer
Sequence		Roland	7	MV8800	1	AutoSubm
HDR		Digidesign	8	ProTools	1	UnlimTrk
PC 37	192.168.100.102	HPTouch		TL52 64x2		Workst. A
PC 38	192.168.100.103	HP		4650		Workst. B
PC 39	192.168.100.104	Sony		CLS30		Workst. C
PC 40	192.168.100.105			xw9300		Workst. D
PC 41	192.168.100.106	HP		xw4400		Workst. E
UnitorATM 8		Emagic		U8-ATM8		MIDI Inter
MIDI		Digidesign		8x8		MIDI Rout
Link		Linksys		WRVS4400		VPN
CDR		HHB		800		CD Recr.
Wave Rec		MOTU		HD192		HDR-Sync
MC		Roland	3	MC-808	15	GrooveBx
MV		Roland	5	8800	14	WaveP
ONYX		Mackie	3	400	3	WaveM A
Ultra		Behrhinger	3	UcurvPro	12	WaveM B
Sample		Roland	5	SP-606	8	WaveM C
Synth A		Yamaha	6	TyroP800	5	Keybrd A
Synth B		Roland	6	G-70	6	Keybrd B
Synth C		Korg	1	Oasys	7	Keybrd C
Synth D		Alesis	1	Fusion6	9	Keybrd D
Synth E		Korg	1		10	Keybrd E
Synth F		Korg	2	M3	11	Keybrd F
Synth G		Yamaha	4	Motiff XS6	13	Keybrd G

Unix Syntax	Command Option
man	Display man pages.
q	Quit man.
passwd	Set or change password.
cd	Change directory.
ls	List contents of directory.
pwd	Print absolute path name of current location.
uname-n	Print host's name.
cal	Display calendar.
date	Display date.
echo	Repeat argument to standard output.
wc	Display line, word & character count in file.
cmp	Compare files.
diff	Display line differences between files.
touch	Create a new, empty file or update time stamp.
tee	Replicate output within pipeline.
mkdir	Create new directory.
public	Code access from a class.
protected	Package level access, super class inherit members.
private	Code access only from same class code.
cp	Copy file.
mv	Moved or rename file.
r	Read
w	Write
chown	Change ownership.
sed	Stream editor.
cat	View file read only.
more	View file one page at a time.
strings	View embedded ASCII text in the executable file.
sort	Sort columnar data.
history	Display history list of last 16 commands.
find	Search directory structure.
grep	Search for character string.
su	Switch user identity.
who	Display logged into system.

jobs	List jobs in shell.
kill	Kill process.
a	Append text to right of cursor.
ftp	Remote file transfer across platforms.
pwd	Print absolute path name of current location.
pkill	Process kill.
A	Append text at end of line.
i	Insert text at cursor position.
O	Insert new text on current line.
rm	Move file.
redir	Removes empty directory.
chmod	Change permissions on the file or directory.
unmask	View or set default permissions value filter.
lpstat	Display status of print job.
spell	Report misspelled words.
cancel	Cancel print job.

Office	1999	2000	2001
Corporate	405	715	624
Satellite A	27	55	42
Satellite B	37	142	84
Satellite C	43	29	27
Satellite D	25	59	46
Satellite E	17	38	21
Satellite F	0	0	70

1/1/2002	125
2/1/2002	680
3/1/2002	700
4/1/2002	680
5/1/2002	655
6/1/2002	525

Command Syntax

Descriptions

Command Syntax	Descriptions
arap	Connects a CLI line into an ARAP connection.
bg	Puts a job in the background.
connect	Connect to LAT
fg	Returns to an established job.
hangup	Disconnects old jobs and resets users CLI connections.
help	Displays help information for user CLI commands.
hosts	Displays the current host table.
jobs	Displays a list of current jobs.
kill	Terminates job.
lock	Locks a port.
netstat	Displays network statistics.
ppp	Converts a CLI port to a PPP interface port.
queue	Displays information about queued HIC requests or removes a HIC.
rlogin	Connects to a host using the rlogin protocol.
services	Displays the LAT services that have been advertised by LAT Nodes.
slip	Converts a CLI port to a SLIP interface port.
stats	Displays annex statistics.
stty	Displays and modifies CLI port parameters.
Telnet	Connects to a host using the Telnet protocol.
tn3270	Connect to IBM VM / CMS or MVS host tn3270 vari Telnet protocol.
who	Displays Annex users.

Telnet Commands:

[?]	Information about one or all of the commands described in the table.
close	Closes connection to the remote host CLI prompt.
display	Total arguments and definitions of the special characters.
Mode	Specifies the input mode.
echo	Specifies whether echoing is performed by the Annex (local_echo).
open	Opens a connection to specified host.

[-r]	Turns off Telnet protocol interpretation.
[-t]	Opens a transparent TCP connection to the specified port.
quit	Close telnet session.
status	Current status of tell Telnet.
send	Sets one or special character sequences to the remote host.
set	Sets the Telnet special characters.
erase	Sets the erased character that send ec command.
interrupt	Sets the interrupt character entered.
crmod	Toggles carriage return mode
binary	Toggles binary mode.
options	Toggles displaying internal Telnet protocol processing.
escape	Telnet escape character used to enter command mode.
kill	Sets line erase character that sends el command of the Telnet session.
localchars	Toggles the local recognition of Telnet special characters.
eof	Sets character sent to host if Telnet is operating in line-by-line mode.
close	Closes connection to remote host and returns to prompt.
ao	The Telnet abort output sequence.

Queue Arguments: **Brackets Indicate Default Values**

[-h]   Displays only the entries originating from the
               hostname.
[-s]   Displays only the entries requesting service.
[-p]   Displays only the entries requesting connection to port.
[-r]   Removes the entry associated with entry_id from the
               queue.
[-v]   Displays the name and port number for queued
               service available.

Telnet Arguments:

[-1]   Directs LFcharacter to the terminal for each CR
               received.
[-r]   Request raw mode. Process data between the terminal
               TCP line mode.

Netstat Arguments:

[-A]   Default information address of associated protocol
               control blocks.
[-a]   State of all sockets including server processes.
[-C]   Contents of the root catche.
[-i]   The hardware interfaces. eg., AppleTalk. SLIPP, PPP.
[-iaport]   Statistics for a specific Annex ARA interface.
[-ip port]   State of PPP interface.
[-iQ]   Interfaced queues.
[iS]   State of hardware interface and additional SLIP
               interface.
[-f]   Filtering statistics.
[-g]   RIP statistics.
[-m]   Statistics for memory buffer allocation.
[-n]   Network addresses as numbers rather than names or
               symbols.
[-R]   Information on rotaries.
[-r]   Routing tables including dial-out routes.
[-ra]   Only AppleTalk route.
[-ri]   Only IP routes.

[-s]	Network protocol statistics, set corret lat_key, option_key.
[-rs]	Routing statistics
[-t]	Default active connection attached device name.
[-z]	Network zone list.

Managing Multiple Jobs:

bg	Places job in background. More then one type placed in background.
fg	Brings job to forground. Only one job at a time.
hangup	Terminates jobs.
jobs	Display list of active jobs.
kill	Terminates a Job.

Company	Licensed Workstation Software						
Microsoft	XP Pro	OfficePro	ExpresST	Publisher	Blend	Enterprise	SharePoint
Emagic	ES1	ESX 24	EVP 73	Soundrv	Logic 5.		
Mackie	RealTime	Tracktion3					
Drawmer	DDX100						
Acuma	TimePack	DelFactor					
Antares	Autotune						
Steinberg	VirtualGtr	CubaseSX	Nuendo 4	Wave 6			
Corel	Draw 10						
Sonic Foundry	Acid 6	Vegas 7					
Metacreations	Headline						
Procreate	Knockout						
MGI	Video	Photo					
Adobe	Illustrate	After Eff	Photoshp	Livemotion	GoLive	Premiere	Press
Adobe	CS3	CS3Pro	Studio 8	Studio MX	VCPro2		
Tascam	GigaStudio		GVI				
DigiDesign	ProTools	HD	Accel 3				
E-Frontier	ANimeST	Poser 7	Shade 8	FigArtist			
MOTU	DP5						
Apple	Logic 8	Motion	Aperture	FinalCut			
Alien Skin	Blowup	Snap Art					
Erain	Swift 3D						
Ableton	Live 7						
Xara	XtremePro						
OnOne	Fractals						

Index

www.ingramcontent.com/pod-product-compliance
Lightning Source LLC
Chambersburg PA
CBHW032011170526
45157CB00002B/643